Packet Based Communications

M000165723

Edited by
Fraidoon Mazda
MPhil DFH CEng FIEE

With specialist contributions

Focal Press
An imprint of Butterworth-Heinemann
Linacre House, Jordan Hill, Oxford OX2 8DP
A division of Reed Educational and Professional Publishing Ltd

R A member of the Reed Elsevier plc group

OXFORD BOSTON JOHANNESBURG
NEW DELHI SINGAPORE MELBOURNE

First published 1996

British Library Cataloguing in Publication Data
Mazda, Fraidoon F
    Packet Based Communications
    I. Title
    621.382

ISBN 02405 1455 6

Library of Congress Cataloguing in Publication
Mazda, Fraidoon F.
    Packet Based Communications/Fraidoon Mazda
    p. cm.
    Includes bibliographical references and index.
    ISBN 02405 1455 6
    1. Telecommunications. I. Title
    TK5101.M37 1993   92-27846
    621.382–dc20   CIP

Printed and bound in Great Britain by
Biddles Ltd, Guildford and King's Lynn

Contents

Preface

Packet based communications systems have traditionally been widely used for the transmission and switching of data traffic. However, with the advent of newer technologies such as Asynchronous Transfer Mode (ATM), packetised voice is also becoming a reality. This book looks at the various aspects of packet based communications and its application within local and wide area networks.

Chapter one introduces one of the basic concepts of packet communications, the OSI seven layer model. Chapter two then describes another of the fundamental requirements of packet communications; the methods used for multiple access to the same transmission medium by several information sources.

The operation of packet switched networks, and its relation to the OSI reference model, is described in Chapter three. This is developed further in Chapter four which covers the principles of fast packet switching technologies, both frame relay and cell relay.

Chapter five describes one of the cell relay technologies, ATM, in further detail. The ATM standard is still developing and this material has been updated to keep pace with changes. ATM promises to be the prime technology of this decade for the communication of voice and data, although it is currently being used primarily for private networks. Chapter six then describes the principles of local and wide area networks and the components which go to operate them.

Eight authors have contributed to this book, all specialists in their field, and the success of the book is largely due to their efforts. The book is also based on selected chapters which were first published in the much larger volume of the *Telecommunications Engineers' Reference Book*, now in its sixth edition.

Fraidoon Mazda
Bishop's Stortford
April 1996

List of contributors

Jim Costello
Telecommunications Consultant
(Chapter 3)

Paul Dyer
Nortel Ltd
(Chapter 3)

Harold C Flots
BSEE MSSM
Omnicom Inc.
(Chapter 1)

Mike Hillyard
BSc (Eng) CEng MIEE
BNR Europe Ltd.
(Chapter 5)

David L Jeans
BNR Europe Ltd.
(Chapter 3)

Fraidoon Mazda
MPhil DFH CEng FIEE
BNR Europe Ltd.
(Chapters 2 and 7)

K L Moran
BSc (Hons) CEng MIEE
Sprint International
(Chapter 4)

Hubert A J Whyte
Newbridge Networks Ltd.
(Chapter 6)

1. The OSI reference model

1.1 Introduction

Effective and meaningful interchange of information is an essential element in the operation of enterprises and the conduct of business activities. The dramatic advances in computer/communications technology are now providing comprehensive capabilities for the establishment and evolution of distributed information systems fully interconnected and integrated to serve as the foundation of business around the world. These distributed information systems have diverse operational requirements and will be supported by continually advancing technologies through a vast world-wide multivendor marketplace of products. The question that immediately arises is: 'How can compatibility among the variety of systems, designs, technologies, and manufactures be realised without constraining innovation, performance, and ongoing evolution?'

In 1978, the International Organization for Standardization (ISO) established a massive standards development programme called Open Systems Interconnection to establish an architecture and family of standards that will serve as the generic basis for compatibility among systems for information interchange.

This chapter presents the perspective, concepts, and functions of the Reference Model for Open Systems Interconnection (OSI), defined in International Standard ISO 7498, and the structure of the comprehensive family of International Standards that have been established for distributed information systems.

1.2 OSI environment

The OSI architecture and family of standards, which specify the services and protocols for interchange of information between systems, have been defined to provide an operating environment for the implementation of distributed information systems from a multivendor market-place. Interchange of information is in digital form and can convey data, voice, and image communications. The interconnected telecommunication resources can be dedicated transmission paths or switched services on a demand basis. Switched paths interconnecting communicating users can be on a fully-reserved basis or on a demand basis using various switching technologies.

An illustrative distributed information system is shown in Figure 1.1. The many different types of systems shown contain Application Processes (APs) that may need to communicate. APs can be manual, computerised, or physical. For example:

1. A manual process could be a person operating a 'point of sale' terminal entering data or receiving an output.
2. A computerised process could be an operational program in a 'host' computer performing its task, such as an accounting program processing a payroll.
3. A physical process could be the operation of a 'robot' in a manufacturing plant.

Communications are often required in order for APs to perform their designated tasks. The data that an AP requires may need to be retrieved from a remotely located data base. The operations that an AP requires may not be supported by its local processing resources, and therefore, it must share additional resources in a remotely located system to accomplish its task. Finally, upon completion of its operations, the AP may need to deliver the resulting data to a remotely-located data base for further processing or analysis.

The interconnecting network shown in Figure 1.1 provides the telecommunication paths for information interchange between the systems. The OSI architecture defines an orderly structure of functionality to facilitate successful communications between APs. There are two basic components to ensure successful communications. The first is the transparent movement of the data between systems. The second is ensuring that the data arriving for the destination AP is in a meaningful form that can be immediately recognised and processed.

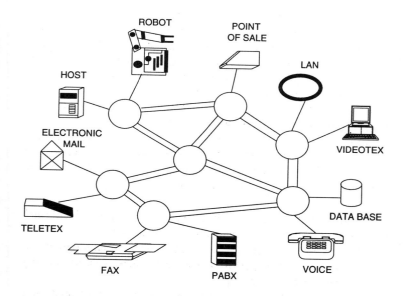

Figure 1.1   Distributed information systems

OSI standards only define those functions that are necessary to facilitate communications between systems.

They do not describe the specific implementation, design, or technology that is used. These are left to the innovation of the system developer.

The functional components of a computer based system are shown in Figure 1.2.

The processing part, information and memory part, and internal communications (comm) part are not constrained by the standards. The OSI elements are additional functions in the communications part that are only invoked when communications with other systems are required. Systems that implement the OSI functions are therefore called Open Systems and can participate in information interchange using the OSI environment.

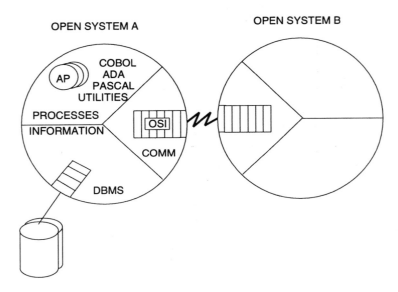

Figure 1.2   Functional components of a computer based system

1.3 Layered architecture

1.3.1 Basic principles

There are many ways to describe and characterise systems for information technology applications. Many arbitrary alternatives could be just as effective, but it is imperative to have as much inherent flexibility as possible in the structure and to have a solution that is globally agreeable. Only then can an architecture be established as a generic basis that will continue to evolve to accommodate advancing technology and expanding operational requirements.

The concept of layering has been widely accepted. Only a minimum number of layers has been defined, thereby keeping the structure basically simple, whilst no individual layer should be so functionally complex that it is unwieldly. The modular structure of

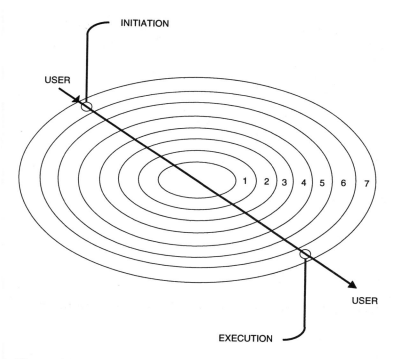

Figure 1.3 Onion skin structure

layering provides a great deal of flexibility by enabling implementations to change over time with new technology, by allowing tailoring of the invoked functions to optimise a specific operating configuration, and by satisfying the broadest range of applications.

A layered architecture, as shown in Figure 1.3, can be referred to as an 'onion skin structure' to describe a communications environment. The users on either side represent the corresponding APs that are interchanging information. In the communication process, the functions of each of the layers (layer 7 through 1) are invoked in the originating system and conveyed to the destination system where they are executed from layer 1 through 7 in the destination system. Very simply, in a point-to-point configuration, each layer is traversed

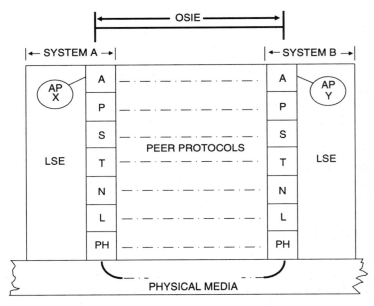

Figure 1.4   OSI environment

twice, firstly for initiation and secondly for execution of the appropriate functions to ensure a successful communication. As interconnecting configurations become more complex with diverse paths through switched telecommunication resources, some of the layers may be traversed multiple times in performing relaying functions to support the information flow between systems.

An overlay of the layered architecture to real systems is shown in Figure 1.4. Each open system has its local system environment (LSE), which includes the basic processing, information, and communication components shown earlier in Figure 1.2. Part of the OSI environment (OSIE) is also shown in each system. The AP resides in the LSE of a system and binds to the OSIE for communicating with another AP in a remotely-located system. Each system has a component of each of the layers that are referred to as layer entities. An instance of a layer entity represents the set of functions of the specific layer that are active in a system to support the instance of communication. The functions are invoked in the layer entities in the originating system and executed by the respective layer entity in the destination system. Specifically, the functions invoked by the entity in layer 7 of system A are executed by the entity in layer 7 of system B. These functions are logically conveyed between systems using peer protocols, which are defined for each of the layers to perform particular tasks. Each of the descending layers in the originating system adds its functions to the layer above as the communication proceeds layer-by-layer. The physical media, which are not part of the OSIE, provide the interconnecting transmission paths between the systems. At the destination system, each layer in turn executes its functions and passes the action to the next layer above, in ascending order.

1.3.2 Layer concept

The concept of a layer is illustrated in Figure 1.5. OSI characterises an (N)-layer as a subdivision of the OSI architecture that interacts with elements in the next layer above, the (N+1)-layer, and the next layer below, the (N-1)-layer. The (N)-entity is an active element within the (N)-layer in a real system, and peer-entities are (N)-entities within the same layer, but in different systems.

The (N)-service relates the collective capability of the (N)-layer and underlying layers provided to support the (N+1)-entity. Associated with each (N)-service are primitive interactions that pass various parameters. The primitives will be described later in this chapter. The semantics of the services and primitive interactions are defined in the family of OSI standards to facilitate an understanding of the dynamics of the OSIE but are not critical to OSI conformance and interoperability between systems.

Peer entities communicate with each other using an (N)-protocol, which is a set of rules and procedures to facilitate an orderly communication in the performance of the associated functions. The peer protocols are the critical specifications of the OSI family of standards that facilitate interoperability between systems and are the basis for determining OSI conformance.

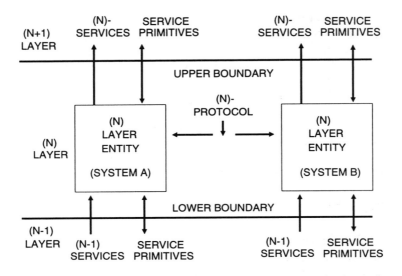

Figure 1.5   Concept of a layer

In the case of the uppermost layer, there is no (N+1)-layer, but this is the point at which the AP binds to the OSIE during an instance of a communication. There is also no (N-1)-layer associated with the lowest layer because that is the point of interface between the OSIE and interconnecting physical transmission media.

1.3.3 Layer services

Whilst each layer performs its functions independently of other layers, interaction between layers is essential to ensure full co-operation in support of the communication. The services are provided to the (N+1)-layer (see Figure 1.5), which is characterised as the (N)-service-user in each system by the companion technical report, ISO TR 8509. This concept is illustrated in Figure 1.6, which applies to any of the layer boundaries. The (N+1)-entity in each system is the service user of the functions of the (N)-layer and any lower layers, which are characterised collectively in Figure 1.6 as the (N)-service-

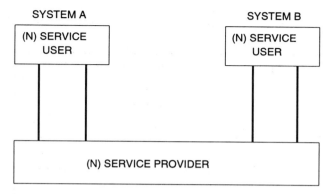

Figure 1.6 Layer service model

provider. Therefore, each of the layer services can be defined from specific vantage points using this model.

Four interaction primitives are defined to further characterise the semantics of the dynamics of operation within a system as follows:

1. REQUEST is provided by the originating service user to activate a particular service by the service provider.
2. INDICATION is provided by the service provider to the destination service user to advise that a particular service is activated.
3. RESPONSE is provided by service user back to the service provider in response to an incoming indication primitive.
4. CONFIRM is returned by the service provider to the originating service user to acknowledge completion of a requested service.

The primitives carry important parameter information between the layers. The nature of the parameters may vary for different services, but they can include address information, data to be transferred, and invocation of service options. The specifics of the services and associated primitives for each layer are covered by the specific layer service definitions, which are part of the OSI family of standards.

1.3.4 Communication between peer entities

Two modes of communication are defined in the OSI Reference Model: connection-mode and connectionless-mode. Connection-mode operation was the original basis for OSI. Later, the connection-less-mode was defined in ISO 7498 AD1.

The connection-mode functions through three phases of operation in support of an instance of communication. These are establishment, data transfer, and release. The initial definition of the OSI Reference Model included only the connection-mode for all the layers.

An (N)-connection represents an association that has been established between (N+1)-entities and establishes the path for information flow that is maintained for the duration of the instance of communication, as shown in Figure 1.7. It is a result of the (N)-layer and the underlying layers collectively. Each layer operating in the connection-mode follows the procedure to establish the association, or connection, with its peer entity. Upon termination of the communication, the connection of each layer is released.

The establishment of a connection allows negotiation of parameter values and options between corresponding entities. It has a distin-

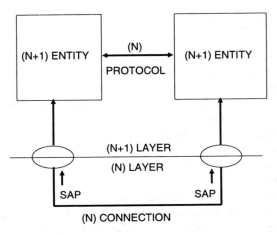

Figure 1.7   (N) connection

guishable lifetime that can support interchange of multiple data units over a period of time.

In contrast, the connectionless-mode does not establish and maintain a relationship between corresponding entities. Each transfer of data in the connectionless-mode is a separate independent action for that layer. It has often been called the 'message mode' or 'datagram mode' where each data unit is routed to the destination on an individual basis with no regard to data units transferred before or after it.

The connectionless-mode simplifies operations where the overhead of error control and recovery is not needed. It also eliminates the overhead of establishment and release of connections for a faster reaction communication. Typically, the connectionless-mode has been applied to routing data to destination systems through telecommunication resources. In this case, some layers are operated in the connectionless-mode, whilst others are operated in the connection-mode. A discussion regarding the use at different layers is included later in this chapter.

In the future, the upper layers directly supporting applications may also use the connectionless-mode for certain operations, such as broadcast to multiple APs.

1.3.5 Relaying

A layer entity can also provide a relaying function to facilitate the routing of data along the path to the destination system. An (N)-relay-entity does not provide any service to an (N+1)-entity but only uses the services from a receiving (N–1)-entity. It processes the incoming data unit and forwards it to the (N'1)-entity associated with the next path for the data to flow through. This concept is illustrated in Figure 1.8. The relay system is identified between the two end systems, which are the source and destination systems for the instance of communication.

A relay system may have a number of possible paths to select from for routing data to the addressed destination, as illustrated by the lines radiating from the ellipse in Figure 1.8. Relay systems apply to both packet and circuit switching and can function in either the connection-mode or connectionless-mode depending upon the specific oper-

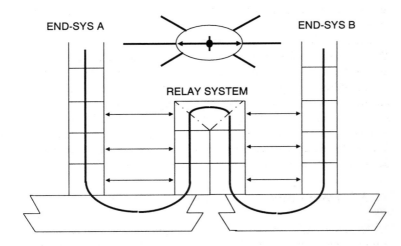

Figure 1.8   Relay function

ational configuration. The circles in Figure 1.1 along the paths interconnecting the end systems are examples of intermediate relay systems that participate in OSIE to support the communications.

1.3.6 Identifiers

The principles of naming and addressing have been included in the
OSI structure to facilitate the mapping of the information flow
through the layers within a system and to find the path between
systems for transmission of the data to the destination system. The
conceptual port between an (N)-layer and an (N+1)-layer is called a
Service Access Point (SAP), which is shown in Figure 1.7. The
location of the SAP is identified by an (N)-address, which may be
globally unique or may only be specific to a particular instance of
communication, depending on the layer involved.

A single SAP can support multiple instances of connections, which
are identified by (N)-connection-endpoint-identifiers. A layer bound-

ary may support more than one SAP between the same or different entities of the (N)-layer and (N+1)-layer.

The provision of (N)-titles is used to uniquely identify individual (N)-entities to facilitate the routing and relaying functions. An (N)-directory is the function that is used to translate a global title into an (N−1)-address to which the (N)-entity is attached. The complex and abstract subject of Naming and Addressing is dealt with in detail in Part 3 to ISO 7498, and the provisions for global Network SAP addressing is defined in the OSI Network Layer Service Definition ISO 8348 AD2. The addressing structure of the upper layers is defined in the Application Layer Structure of ISO 9545.

1.3.7 Data units

As discussed earlier in this chapter, a communication activity from an AP in a system progresses from layer to layer as the required functions are initiated to support the communication. In turn, at the destination system, the incoming communication ascends through the layers as the functions are executed to provide a successful communication to the destination AP.

This transfer of information between the systems involves the construction and forwarding of control information and data units that convey the actions and substance of the communication. Figure 1.9 presents the basic data unit structure that passes layer-by-layer within a system. A service data unit (SDU) is the data that is passed to the (N)-layer from the (N+1)-layer for forwarding to the peer (N+1)-entity in the destination system. Therefore, the (N)-layer is responsible for transferring the SDU transparently, so that it is delivered to the destination (N+1)-entity exactly as it was received by the (N)-entity in the source system.

The SDU is received by the (N)-entity as a parameter of a request primitive issued by the (N+1)-entity. The (N)-entity then produces the protocol control information (PCI), which initiates the required functions and provides the appropriate parameter information for execution by the peer (N)-entity in the destination system.

The combination of the SDU and PCI becomes the protocol data unit (PDU) for transfer to the peer (N)-entity via the (N−1)-connec-

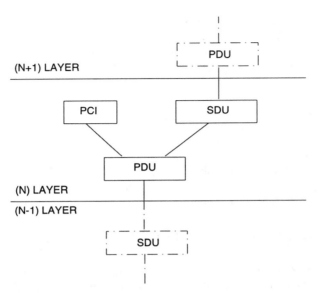

Figure 1.9 Data units

tion or connectionless path, depending upon the mode of operation. The (N)-PDU becomes the (N−1)-SDU, which is them processed in a similar manner with the addition of a (N−1)-PCI to produce the (N−1)-PDU for the next layer below. Recursively, this process occurs in each descending layer until the data leaves the OSIE in transmission on the physical media.

At the destination system, the PDU enters a layer as a parameter of an indication primitive. The layer entity processes the PCI part, executes the functions, and discards the PCI, leaving the SDU to pass to the next ascending layer via an indication primitive. Again, each layer in turn recursively processes the data units until the final data is delivered to the destination AP in a form that is meaningful for processing in the local environment of the destination system.

The layer protocols in the family of OSI standards support the many functions in the different layers. The protocols define the peer entity interactions, along with the syntax and information content of the PCIs that convey the actions between communicating systems. These protocol standards are the specifications that provide the compatibility and interoperability between systems in the world-wide multivendor market-place of products for information technology applications.

1.4 Additional layer elements

The OSI Reference Model defines many additional layer elements that further characterise the OSIE. The various elements defined can apply at some or all layers, depending on the specific service requirements for the particular layer.

1.4.1 Multiplexing and splitting

The layer entities have been shown as mapping on a one-to-one basis from layer to layer. However, additional flexibility is provided through the mapping of multiple entities to provide for more effective use of resources as needed. Multiplexing involves the mapping of multiple (N)-connections to a single (N–1)-connection. This configuration enables bandwidth sharing of the connection to support multiple communications. The demultiplexing is done in the respective peer entity in the destination system.

Splitting involves the mapping of a single (N)-connection to multiple (N–1)-connections. This configuration enables expansion of the bandwidth to support a communication that requires a higher throughput of data than can be supported by the bandwidth of a single (N–1)-connection. Splitting can also apply to high survivability situations where the loss of a sole (N–1)-connection would be seriously disruptive to an operation, whereas the loss of one of many connections in a split configuration would only reduce the bandwidth.

1.4.2 Data transfer

There are two basic types of data transfer: normal and expedited. The unit of transfer by a layer is its protocol data unit (PDU). Every layer has the function for transfer of data in support of the instance of communication.

1.4.2.1 *Connection-mode*

In the connection-mode, normal data transfer is accomplished by a layer through an (N–1)-connection that has been established with the destination (N)-entity. Some of the layers will also employ the following additional functions:

1. Flow control regulates the rate of data transfer, so that PDUs do not arrive at the destination (N)-entity faster than they can be processed. Peer-to-peer protocol flow control mechanisms are generally employed to co-ordinate the rate of transfer between the peer (N)-entities. Flow control may also be applied across the layer interface between an (N+1)-entity and an (N)-entity.

2. Segmenting enables a single (N)-SDU to be broken up into smaller units for processing into multiple (N)-PDUs. At the destination (N)-entity, the SDU segments are reassembled into a single (N)-SDU before passing to the (N+1)-layer. An (N)-protocol mechanism is employed to keep track of the segments during transfer so they can be recombined at the destination.

3. Blocking enables multiple (N)-SDUs to be mapped into a single (N)-PDU. Although this function is specified in the OSI Reference Model, it has not been employed in any of the OSI protocols or configurations.

4. Concatenation enables multiple (N)-PDUs to be mapped to a single (N)-SDU in the (N–1)-layer.

5. Sequencing returns data units to the same order at the destination (N)-entity that they were sent in from the originating (N)-entity. This situation could particularly occur in connectionless-mode data transfer where each data unit could take a different route to the destination system. The (N)-protocol PCI contains a sequence number that is used for the resequencing function.

6. Acknowledgement provides confirmation for delivery or receipt of PDUs by a destination (N)-entity. Generally, the (N)-protocol PCI provides the acknowledgement mechanism and sequence numbers to facilitate this function, which generally provides for recovery of lost data units.

7. Error detection and notification identifies corruption of data during the transfer process between (N)-entities. Recovery from such errors can be accomplished by the same mechanism used for lost data units found by the acknowledgement function.

8. Reset provides for reinitialisation of an (N)-connection without the loss of associations and connections of the upper layers.

9. Routing provides for relaying PDUs across multiple (N)-entities for establishing a connection and transferring data, or for connectionless-mode data transfer.

The connection-mode data transfer also provides for expedited data, which is not intended to be used on a regular basis. Generally, expedited data is limited in amount of data transferred and its frequency of transfer. It is not subject to the same flow-control constraints, if any, as normal data transfer. The basic rule is that an expedited data unit should not arrive at the destination any later than a normal data unit sent after the expedited data unit. In other words, it should not be any slower than normal data but may be faster. Expedited data service is invoked as required by the upper layers. Expedited data may also be acknowledged by the destination (N)-entity.

1.4.2.2 *Connectionless-mode*

Data transfer in the connectionless-mode involves minimal functionality. Data units are sent, routed, and delivered with no knowledge being maintained by the (N)-layer of their status along the way. Some have referred to this mode as 'send and pray' because there is no assurance that delivery has been successful. On the other hand, the shortcomings of this simple approach can be compensated for by use of appropriate functions at higher layers, which are operating in the connection-mode.

1.5 Specific OSI layers

As explained earlier in this chapter, the number and functional configuration of the specific layers for the OSIE are somewhat arbitrary. A number of arrangements to satisfy the requirements are possible. However, the agreement has been made internationally for the OSI configuration, which provides a solid generic basis to characterise distributed information systems whilst establishing a structure for accommodating advances in technology and expanding operational requirements through continuing evolution.

The required functions to support successful communications between APs are distributed among seven architectural layers. The layers, as shown in Figure 1.10, are defined below.

In summary, the upper three layers provide the functions for a meaningful communication so that information reaching the destination AP is processible and compatible with the local systems environment. The lower four layers provide the connectivity for interchange of data between systems and collectively represent the 'bit pipe' for transparent data transfer.

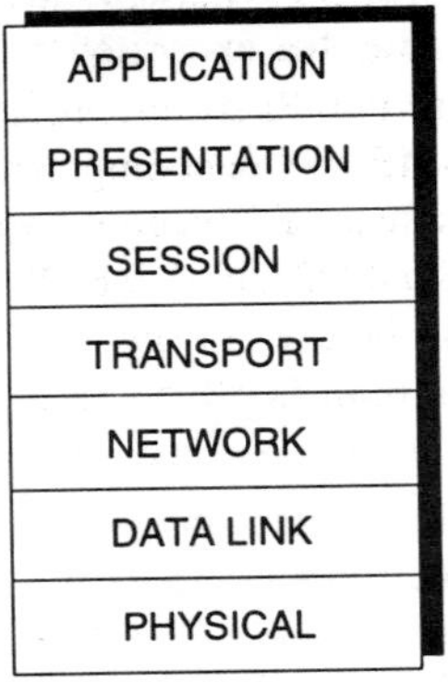

| APPLICATION |
| PRESENTATION |
| SESSION |
| TRANSPORT |
| NETWORK |
| DATA LINK |
| PHYSICAL |

Figure 1.10   Layers of the OSI model

1.5.1 Application layer

The structure of the application layer has changed from that defined in the OSI Reference Model of ISO 7498. The current concept defines the application layer entity as consisting of two types of service elements. The Association Control Service Element (ACSE) of ISO 8649 and ISO 8650 establishes the association between communicating APs and sets the context of the communication. The context refers to the other operational elements that are needed to support the specific nature of the communication. These are defined as the Application Service Elements (ASEs), which may be used individually or in multiples to satisfy specific operational contexts. Example ASEs include: File Transfer, Access, and Management (ISO 8571); Virtual Terminal (ISO 9040 and ISO 9041); Transaction Processing (ISO 10025); Commitment, Concurrency, and Recovery (ISO 9804 and ISO 9805); Remote Operations (ISO 9072); Reliable Transfer (ISO 9066); Remote Database Access (ISO 9579); and others. There is a richness of functionality to select from in the application layer.

1.5.2 Presentation layer

The application layer is defined in terms of its semantics and described in terms of an abstract syntax of the data types involved. The Abstract Syntax Notation No. 1 (ASN.1) of ISO 8824 has been widely accepted for use in OSI specifications. When data are transferred through the OSIE, an encoding agreed upon by the peer presentation entities is needed. The common presentation layer service and protocol of ISO 8822 and ISO 8823 are used to establish the transfer syntax that will be used to support the communication. The Basic Encoding Rules of ISO 8825 are now widely used in OSI applications, but other transfer syntaxes can also be selected.

1.5.3 Session layer

The session layer facilitates orderliness in a communication. A session connection is established, and the specific functions that are to be used are selected. Among the choices in typical use are: use of either two-way simultaneous or two-way alternate data transfer, expedited data transfer, synchronisation and recovery of data, activity management, and exception reporting. The session service and protocol are specified in ISO 8326 and ISO 8327, respectively. The connectionless-mode in the upper three layers is under development.

1.5.4 Transport Layer

The transport layer provides for the transparent transfer of data between peer session layer entities and ensures that the appropriate quality of service, which was requested by the originating session entity, is maintained. In this regard, the transport layer is required to select the network layer service that most closely matches the requested requirement. In addition, it may also have to invoke additional functionality to ensure the proper quality of service. Five classes of connection-mode protocol have been defined in ISO 8073 to accommodate the variety of operational conditions. The transport service is defined in ISO 8072. There is also a connectionless- mode transport protocol specified in ISO 8602.

1.5.5 Network layer

The network layer provides the functions for the telecommunication resources that provide the paths for transfer of data between systems. Real telecommunication services, such as may be provided by a public data network, are defined by OSI as subnetworks. Paths between systems may involve switched services and interconnection of multiple subnetworks en route. Both connection-mode and connectionless-mode operations have been widely applied in the network layer. The Network Service Definition is provided in ISO 8348 and a number of standards are available for the various protocols and configurations that can be employed.

1.5.6 Data link layer

The data link layer provides the means for synchronising the bit stream flowing to and from the physical layer and for detection of errors due to transmission problems. Typically, Local Area Networks (LANs) operate the data link in connectionless-mode, whilst wide area network applications operate in the connection-mode. In the connection-mode, the additional functions of flow control and error recovery are included. The Data Link Layer Service Definition is specified in ISO 8886.

1.5.7 Physical layer

The physical layer provides the functions for transparent transmission of the bit stream between data link entities. When the physical layer is active, the bits flow through to the transmission medium. The electrical characteristics and physical connector serve as the interface between the OSIE and the external transmission media. The Physical Layer Service Definition is specified in ISO 10022. A number of physical interfaces are available for a variety of transmission environments.

1.6 Additional provisions

Three additional parts of the OSI Reference Model have been published that cover some other very important aspects of distributed information systems.

These are summarised as follows:

1.   ISO 7498-2: Security Architecture provides the framework for development of various security provisions to protect system integrity and unauthorised access to user data.
2.   ISO 7498-3: Naming and Addressing provides the concepts and principles for identifying objects through naming and location of objects through addressing mechanisms.
3.   ISO 7498-4: Management Framework provides the concepts and principles for management of the resources of the OSIE to ensure continuing and effective operation.

1.7 Conclusions

The OSI architecture establishes a generic foundation for implementation and evolution of distributed information systems to support a diversity of operational requirements. Whilst the richness of functionality that is available can lead to significant complexity, not all the capabilities are needed for every application. It is important to select only those functions that are needed for particular operational configurations, and no more. Therefore, the complexity will vary as needed to fulfil specific requirements.

OSI is a living architecture that will facilitate evolution to new technologies and operational requirements, but the basic structure must be preserved to ensure an orderly transition on a common basis. New functions may be added to layers in the future, whilst some of the existing functions will become disused. Even some layers may be inactive in certain applications. Nevertheless, keeping the basic OSI structure intact will ensure a consistent evolution in the future.

1.8 References

Abramowicz, H. and Lindberg, A. (1989) OSI for telecommunication applications, *Ericsson Review*, (1).

Smythe, C. (1990) Networks and their protocols, *Electronics & Communication Engineering Journal*, February.

Whitty, T. (1990) OSI: the UK approach, *Communications*, May.

Wood, B. (1988) Standards for OSI — present status, future plans, *Telecommunications*, March.

2. Multiple access techniques

2.1 Introduction

Communication bandwidth is a scarce commodity, whether it be carried on copper wire, fibre optic cable or by the air waves. The need to share this bandwidth fairly and efficiently between multiple users is therefore crucial. Many techniques have been developed to enable multi-users to share, and especially to access, the same communication channel, and these have all been employed in various applications. Each technique has advantages in certain environments, but often different techniques are used for the same or similar application. For example, in digital cordless telephone, CT2 uses FDMA whereas DECT and CT3 are based on the TDMA. Also, in the UK, personal communication networks (PCN) are based on TDMA multiple access techniques, whereas the system proposed by Nynex in the USA is to use CDMA (Ketseoglou and Zimmerman, 1994).

This chapter describes and compares the various multiple access techniques and introduces a classification system for the many systems which exist. No reference will be made to any particular application, these being covered in other chapters. Differences between mobile and fixed systems are also narrowing and there are many ways in which mobile communications occur, for example wireless local and wide area networks, PMR, cellular, PCN and satellite.

Network topology sometimes plays a role in the selection of the multiple access technique. Figure 2.1 shows the most commonly used forms. The star network is usually seen in configurations such as lines terminating within a central controller or switch. It is well suited to carry out a polling mechanism from the central location to outlying nodes, for multiple access.

Ring topologies are often preferred since they are not dependent on a single control site, and can generate traffic in either direction

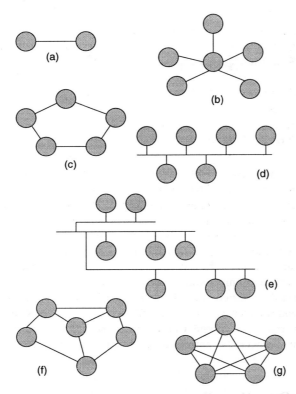

Figure 2.1   Popular network topologies: (a) point-to-point;
(b) star; (c) ring or loop; (d) bus; (e) tree; (f) mesh;
(g) fully interconnected

around the ring, so providing alternate transmission paths and therefore survivability. The ring architecture is suited to multiple access methods such as token passing and slotted ring, as described later.

The bus structure provides simultaneous access to a central bus by several nodes, and this is the most commonly used structure within local area networks of office buildings, with CSMA/CD (Ethernet) the most popular multiple access mechanism. Tree topologies can be considered to be generalised forms of bus structures and have similar characteristics.

Another consideration in the choice of a multiple access technique is the geographical area covered by the network. Local Area Networks (LAN) spread over a few kilometres, usually an industrial site or university campus, and are owned by a single organisation. They can operate at high data rates.

Wide Area Networks (WAN) cover a much larger geographical area and are usually composed of LANs interconnected by other communication links hired from a PTT or common carrier. Their operating speed is usually restricted by these links.

Metropolitan Area Networks (MANs) fall in-between LANs and WANs in geographical area and characteristics.

Although multiple access techniques can be used irrespective of the transmission technology, their performance is often different. For example radio networks have a relatively low propagation time, of the order of a few milliseconds. They are largely omnidirectional and are usually sent as a broadcast to all receivers, although some may be out of reach or screened from the source. Satellite based radio systems broadcast their information, but can be made to have a relatively small footprint on the earth. They suffer from long delays, of the order of a quarter of a second (up-link and down-link), which would seriously hamper the performance of some multiple access techniques.

2.2 Queuing theory

Queuing theory is covered in Kleinrock (1976). The basic parameters are introduced in this section.

Queuing analysis usually refers to queues by a notation such as A/B/m. This applies for an infinite population of users, which is usually the assumption made when analysing multiple access techniques.

In this notation A represents the interarrival time probability density, i.e. the time between consecutive arrivals into the queue. B is the service time probability density, i.e. the time required to serve an arrival. The letter m represents the number of servers which are present to accommodate the arrivals.

Values of A and B vary from M (representing Markov or exponential probability density); D (representing Deterministic i.e. all arrivals have the same value); and G (representing General or arbitrary probability density).

The most common is the M/M/1 queue, and this is used in the analysis in this chapter. The assumption is also made that no arrivals are lost, i.e. leave or are rejected from the queue because of long waiting times. Arrivals are also assumed to be served on a first come first served basis, unlike, for example, a hospital queue where the most urgent cases are served first.

If λ is the arrival rate in number per unit interval of time and μ is the service rate in number per unit time, then the traffic intensity ρ is given by Equation 2.1.

$$\rho = \frac{\lambda}{\mu} \tag{2.1}$$

The average number (N) in the queue is given by Equation 2.2.

$$N = \frac{\rho}{1 - \rho} \tag{2.2}$$

As the traffic intensity grows, i.e. the arrival rate approaches the service rate, the value of the traffic intensity factor ρ approaches unity, and the number in the queue rapidly increases towards infinity. The queue is now said to be unstable.

The total waiting time (T_w) in the queue for any arrival is equal to the time before service begins and the actual service time. It is equal to the time between an arrival into the end of the queue and its departure from the queue, with service having been completed. It is given by Equation 2.3 and, for a transmission channel capacity of C bits per unit time, it may be rewritten as in Equation 2.4.

$$T_w = \frac{N}{\lambda} = \frac{1}{\mu - \lambda} \tag{2.3}$$

$$T_w = \frac{1}{\mu C - \lambda} \qquad (2.4)$$

2.3 Performance parameters

This section introduces some of the concepts which need to be considered when comparing the performance of different multiple access techniques. These will then be used in the following sections which describe the methods in more detail.

The key performance measures are:

1.   The elapsed time between a packet of information being ready for transmission at a transmitting node and when it is successfully received at the receiving node. This is referred to as the delay and is usually denoted by D.
2.   The total rate of information flow, the carried load, between the nodes on the network. This is referred to as the throughput, S. As will be seen later, throughput is also a measure of utilisation.
3.   The fraction of the total capacity of the transmission medium which is used to carry information between nodes. This is referred to as the utilisation, U.

Several parameters need to be considered when defining the performance of a network. These are:

1.   The number of nodes (N) on the network.
2.   The average arrival rate (λ) at each node, assumed to have a Poisson distribution and to be the same for all nodes. This is also the rate of data generated by each node.
3.   The capacity (C) of the transmission medium usually specified in bit/s, and also referred to as its bandwidth.
4.   The velocity of propagation (v) within the medium. For example, for coaxial cable this is about 65% the velocity of light, or 2×10^8 m/s.
5.   The average length (L) of the message frame in bits.

6.   The offered traffic (G), i.e. the total rate of information presented to a network for transmission. Whereas the average arrival rate is a measure of the data traffic, the offered traffic includes items such as control data (e.g. acknowledgements), packets destroyed due to collision, retransmissions, etc.

7.   The length (l) of the communication path.

Generally the throughput (S) and offered traffic (G) are normalised, i.e. expressed as a fraction of the capacity of the transmission medium. For example, if the successful transmission between nodes totals 10Mbit/s, and the capacity of the medium is 100Mbit/s, then the throughput S is stated to be 10/100 or 0.1. Therefore S is also a measure of the utilisation of the medium, as stated earlier. For stability the steady state arrival rate at a network must not exceed the rate at which this data can be transmitted over the network, otherwise the queue would grow without control.

Under ideal conditions the relationship between the throughput (S) and offered traffic (G) is as in Figure 2.2. The traffic throughput increases linearly, as more traffic is presented to the network, until it reaches the maximum capacity of the transmission medium, at which stage the throughput flattens out.

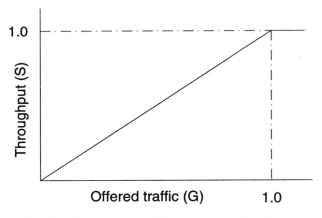

Figure 2.2   Idealised curve of S/G or medium utilisation

The length of the medium in bits is give by Equation 2.5 and it is a measure of the number of bits which are in transit between communicating nodes at any time.

$$l_{bit} = \frac{C\,l}{v} \tag{2.5}$$

A very important dimensionless parameter used in the definition of the performance of multiple access systems is denoted by α as in Equation 2.6.

$$\alpha = \frac{\textit{Length of medium between nodes (in bits)}}{\textit{Length of data frame (in bits)}} \tag{2.6}$$

Using Equation 2.5 gives the value of α as in Equation 2.7.

$$\alpha = \frac{\dfrac{C\,l}{v}}{L} = \frac{\dfrac{l}{v}}{\dfrac{L}{C}}$$

$$= \frac{\textit{Propagation time in the medium}}{\textit{Transmission time of the message}} \tag{2.7}$$

For satellite systems α is usually large, of the order of 10–100, due to the round trip delays, whilst for terrestrial systems it is normally relatively small, usually less than 1.

The utilisation factor (U) is therefore given by Equation 2.8, and the theoretical curve of Figure 2.2 can be modified to take factor α into account, as in Figure 2.3.

$$U = \frac{\textit{Total information flow or throughput (S)}}{\textit{Total capacity of the medium (C)}}$$

$$= \frac{1}{1 + \alpha} \tag{2.8}$$

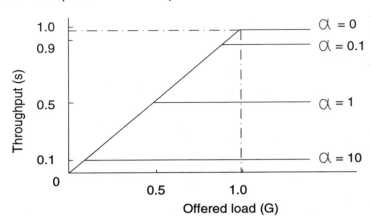

Figure 2.3   Effect of α on the idealised curve of Figure 1.2

2.4 Access systems classification

Figure 2.4 provides a classification for the many different multiple access techniques which have been developed and gives a few examples, described later in this chapter. In pure contention systems a source transmits its data independently of other users, i.e. does not take any action to avoid collision with data being transmitted by other users. If collisions occur these are sensed and the data is retransmitted.

In contention minimisation systems the source takes some action to minimise the occurrence of collisions but does not avoid these altogether, whilst in non-contention systems each source takes due regard of all other sources transmitting on the common line, so that data collisions are avoided. These three major classifications can be subdivided further, as described in the following sections.

2.5 Pure contention systems

The pure contention multiple access techniques are the simplest to implement since a source having data to transmit does so without

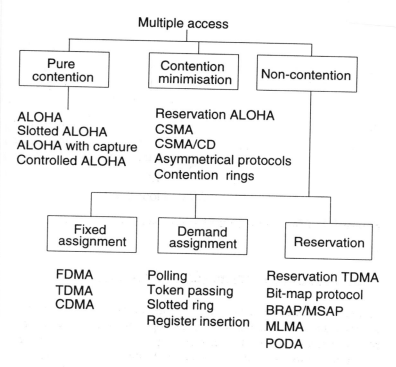

Figure 2.4 Classification of multiple access techniques

regard for any other user of the common transmission medium. Very little sensing or synchronisation mechanism is required, the prime requirement is for an acknowledgement system which can detect when the data has not been correctly received, usually due to a collision with information transmitted by another source, so that it can be retransmitted.

In contention protocols any overhead involved in assigning channel access is independent of the number of users, but depends instead on the channel traffic. This system is most efficient when there are a large number of users, each sending small amounts of 'bursty' traffic.

2.5.1 Pure ALOHA

The pure ALOHA multiple access system (usually referred to simply as ALOHA) was developed by the University of Hawaii for use with satellite based transmission systems. Its operation can be explained by the simplified flow diagram of Figure 2.5. Whenever a data source has information to send, usually a packet of data, it does so without regard for any other data source which may also be using the transmission medium. It then waits the appropriate amount of time for an acknowledgement. If this is not obtained the source assumes that a collision has occurred. It then waits for a random amount of time before retransmitting the same information.

A random wait or back-off time is required to ensure that two sources, which have just collided, do not again collide when they retransmit their data. Many algorithms are in use for calculating this random time. Also the acknowledgement process varies depending

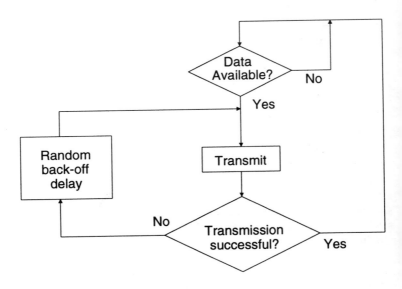

Figure 2.5   Flow diagram of a pure ALOHA access system

on the transmission system being used. For broadcast systems, such as satellites, the sender listens for its message to be broadcast from the satellite, and if it does not occur within a given time (allowing for up-link, down-link and processing delays) then it assumes a collision has occurred.

Figure 2.6 illustrates the process of packet collision within ALOHA. If any parts of two packets overlap (e.g. the first bit of one and the last bit of another) then they are both destroyed. This means that there is a vulnerable period equal to the the sum of the two packet lengths (or twice a packet length, if all packets have the same length) when collision can occur.

In Figure 2.6 packets from users B and D are seen to overlap and collide. After this has occurred it is assumed that there is a finite time t which both senders take to detect that a collision has occurred. They then wait random periods of time (T_1 and T_2) before retransmitting their data, which in this case does not collide with any other data on the line.

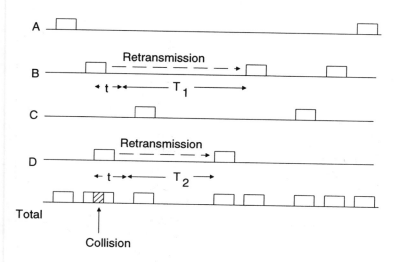

Figure 2.6 Illustration of packet collision and retransmission in ALOHA

The data which is transmitted from the source and eventually reaches its destination is the throughput S. The total traffic on the line, which includes all retransmissions, is the offered traffic G, both usually being normalised to the channel capacity, as stated earlier. Also, as seen from Figure 2.6, collisions and retransmissions add to the delay between the transmission of a packet and it being correctly received at the source.

The performance of an ALOHA system has been analysed by many researchers (e.g. Abramson, 1977; Binder et al., 1975a). It will be assumed that there is a very large number of users (infinite) each sending small amounts of data, in the form of packets, with a Poisson distribution and having a mean of S packets per unit time. For stability $0 < S < 1$. The offered load (G) is also assumed to have a Poisson distribution and $G \geq S$, with $G \rightarrow S$ when the load is low and there are few collisions. It is also assumed that all users are equal, i.e. they have equal probability of transmission.

The performance of an ALOHA channel is given by Equation 2.9 and is plotted in Figure 2.7.

$$S = G\,e^{-2G} \tag{2.9}$$

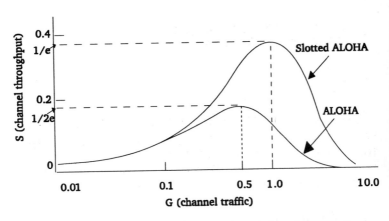

Figure 2.7 Throughput in pure ALOHA and slotted ALOHA channels

From this it is seen that the maximum throughput of an ALOHA channel occurs when $G = 0.5$ and is equal to 0.184, i.e. the channel utilisation is limited to 18.4%. This is the maximum possible utilisation since, as seen later, the delay in packet transmission reaches unacceptable values as this theoretical utilisation figure is approached.

If the offered load exceeds 0.5 the number of collisions increases at a rapid rate, so that although the offered load is increasing this is made up of retransmitted traffic and the useful traffic, or throughput, actually decreases, until it eventually falls to zero.

Although ALOHA has low utilisation it is frequently used for systems which have many users who are transmitting low volume bursty traffic.

If the average length of a message is equal to T_m seconds, the average propagation delay in the medium is T_p seconds, and the average number of retransmissions required before a message is successfully transmitted is n_r, then these are given by Equations 2.10 to 2.12.

$$T_m = \frac{Message\ length\ in\ bits}{Channel\ capacity\ in\ bits\ per\ second} = \frac{L}{C} \tag{2.10}$$

$$T_p = \frac{Length\ of\ medium\ (m)}{Velocity\ in\ medium\ (m/s)} = \frac{l}{v} \tag{2.11}$$

$$n_r = \frac{G}{S} \tag{2.12}$$

The random back-off delay is usually an integral number of packets, if fixed length packets are being used. If this varies between 1 and x packets, then the average back-off delay (T_b) is given by Equation 2.13, and the overall delay or waiting time (T_w) in the communication channel is given by Equation 2.14.

$$T_b = \frac{1 + x}{2} \tag{2.13}$$

$$T_w = (T_m + T_p) + (T_m + T_p + T_b) \, n_r \qquad (2.14)$$

In Equation 2.14 the first bracketed term is the delay experienced due to the initial transmission and would be the overall delay if no collisions were to occur. The second bracketed term accounts for the delays due to collisions. Delay is also often normalised to the packet length in seconds (which is also the message transmission time), and is denoted by D, as in Equation 2.15.

$$Delay \, D = \frac{T_w}{T_m} \qquad (2.15)$$

Figure 2.8 shows the plot of delay for a system operating over channels with significant and variable delay. This also indicates the

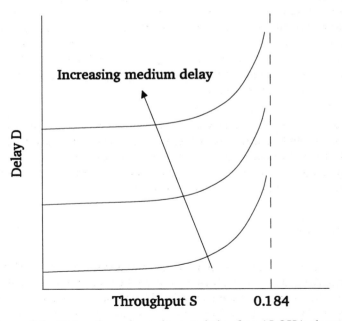

Figure 2.8   Delay-throughput characteristic of an ALOHA channel

very large delays which are experienced as the theoretical maximum throughput of the ALOHA channel is approached.

In order to increase throughput the retransmission should be randomised over longer intervals, to reduce the probability of collisions on retransmission. This means that the value of x, in Equation 2.13, should be large. However, as seen from Equations 2.14, this would also result in increased delay, so that design of multiple access systems often requires a trade off between throughput and delay.

2.5.2 Slotted ALOHA

In the pure ALOHA system collisions will occur whenever any part of one packet overlaps any part of another packet. The vulnerable period is therefore equal to the sum of their two packet lengths, or to two packet lengths if all packets are the same size. An alternative version of ALOHA was developed (Roberts, 1972; Kleinrock and Lam, 1973) which uses fixed length packets and synchronises all transmissions to the start time of the packet. The flow diagram for this system is as in Figure 2.5 except that transmissions can only occur at the start of a packet period.

The disadvantage of slotted ALOHA is the added complexity of requiring a method for synchronising all sources to the start of a packet frame, and the use of fixed length packets. The advantage is that the vulnerable period is now halved, i.e. is equal to the length of one packet. Assuming that all packets are full during transmission, then the throughput is given by Equation 2.16.

$$S = G\,e^{-G} \tag{2.16}$$

Figure 2.7 shows the plot of the throughput curve for a slotted ALOHA channel. It has twice the theoretical maximum throughput (0.368) compared to the pure ALOHA channel and this occurs at a value of G equal to one. This however assumes that all the packets are full, since if they are only half full then the actual throughput is the same as for pure ALOHA.

So far the discussions have assumed an infinite number of users. If they are finite, and equal to N, then the throughput for the *ith* user (S_i) is given by Equation 2.17.

$$S_i = G_i \prod_{j \neq i} (1 - G_j) \tag{2.17}$$

For N identical users, this reduces to Equation 2.18.

$$S = G \left(1 - \frac{G}{N} \right)^{N-1} \tag{2.18}$$

As N tends towards a very large number of users, S approaches the value given in Equation 2.16.

In a slotted ALOHA channel each time a user has information to transmit it must wait for the start of a slot before doing so. This applies also to retransmitted data. There is therefore, on average, an extra half packet time delay compared to pure ALOHA. At the theoretical peak transmission point (G = 1 and S = 0.368) there are an average of 2.7 (= G/S) transmissions for every successful transmission, resulting in an extra delay t_d given by Equation 2.18.

$$t_d = 2.7 \times \frac{Packet\ length}{2} \tag{2.18}$$

Figure 2.9 compares the delay curves between a typical pure ALOHA and slotted ALOHA system, where it is seen that for low throughput (few users and low traffic) pure ALOHA has a lower delay.

2.5.3 ALOHA with capture

ALOHA with capture is a modification to the basic slotted ALOHA system (Roberts, 1975) which works on the principle that if two signals have different strengths, then on collision the stronger one will be 'captured' by the receiver and will get through, so that only

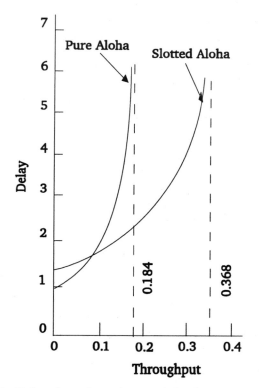

Figure 2.9   Delay-throughput characteristics for pure ALOHA and slotted ALOHA

the weaker signal needs to be retransmitted. This, in effect, reduces the retransmissions by half, and gives almost a 50% increase in throughput over slotted ALOHA, as seen by Equation 2.19. It is assumed that if three stations collide none of them gets through, i.e. no one station is strong enough to dominate any two others.

$$S = G\,e^{-G}\left(1 + \frac{G}{n}\right) \qquad (2.19)$$

In this equation n equals 2 if, for the colliding packets, one or the other gets through; n equals 3 if one gets through a third of the time, the other gets a third of the time, and neither gets through at the other times. For n equal to 2 the value of S is about 50% greater than that attained with a slotted ALOHA system, as stated earlier.

Figure 2.10 compares the performance of ALOHA with capture with pure ALOHA and slotted ALOHA.

Many techniques exist for allocating signal strength to users. For example this could depend on the priority of the user, or could be allocated on a random or pseudo-random basis.

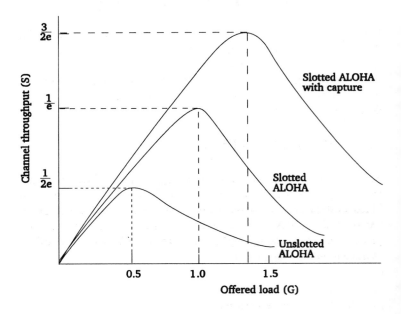

Figure 2.10 Comparison of throughput for pure ALOHA, slotted ALOHA and ALOHA with capture

2.5.4 Controlled ALOHA

An ALOHA channel can rapidly move into congestion if the traffic approaches the theoretical maximum, due to collisions and retransmissions. Several methods have been devised (Lam and Kleinrock, 1975) for controlling the traffic so as to reduce congestion. The prime methods are to reduce the probability of retransmission of collided packets in a time slot, so reducing G, and to deny certain users the right to transmit over a period of time, which in effect reduces the number of users.

2.6 Contention minimisation systems

In pure contention systems a user takes no account of other users, and therefore transmits when it has any data available. There are however several multiple access protocols which, although basically of the contention type, attempt to minimise the amount of contention between users. Some of these contention minimisation systems are described in this section.

2.6.1 Reservation ALOHA

This multiple access method is very similar to the ALOHA (pure and slotted) systems described in the previous section, but uses some form of reservation of capacity as a means of minimising contention. There are several variants of these as described below, and they can be grouped as those needing implicit reservations and those with explicit reservations.

2.6.1.1 *Implicit reservations*

In one such system (Binder, 1975b) the transmission slots are grouped into frames, the number of slots exceeding the number of users. Each user is nominated as being the owner of a slot position within the frame. (Frame 0 in Figure 2.11.)

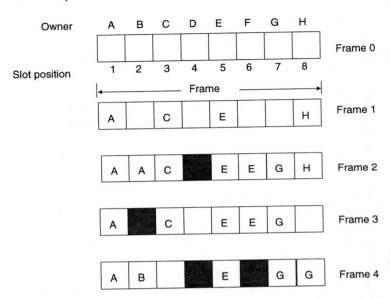

Figure 2.11   Illustration of an implicit reservation ALOHA system

Each user has priority for transmission on his slot. If there are more slots than users then all users contend for the spare slots, using ALOHA or slotted ALOHA. Whilst an owner is transmitting on his allocated slot no other user can interfere. However, if a slot falls idle for a frame, then in the next frame other users can contend for its use. Once a slot has been so seized only the owner is allowed to interfere with the new user. The owner gets his slot back by transmitting on it. A contention occurs and in the next frame only the owner is allowed to transmit.

In Figure 2.11 in Frame 1 owners A, C, E and H transmit on their allocated slots, the other being free. In the following frame (Frame 2) user A and E successfully seize another slot each, whilst G commences transmission. In addition a collision occurs in slot 4, either due to the owner and another station or two other stations simultaneously transmitting.

In Frame 3 slot 4 is vacant, which means that the owner was not involved in the collision in Frame 2, so each station had to wait to make sure. Also a collision has occurred in slot 2, which is due to the owner wanting his slot back.

In Frame 4 the owner B commences transmission in his slot. The two users who collided in slot 4 during Frame 2 operate an appropriate back-off algorithm and will try again. In the meantime other users have decided to contend for slot 4 (one of these may be the owner of the slot) so again contention occurs. In addition the owner of slot 6 has forced a contention to regain his slot.

From this illustration it is seen that the reservation protocol gives high efficiency when users have stream type traffic, with owners transmitting on their slots. However, for bursty traffic, there will be many collisions, and since all users (apart from the owner) need to wait for a frame following a collision, slots can go empty (such as slot 4 in Figure 2.11).

Slots must also remain vacant for one frame following the end of transmission from its owner, although this can be overcome by having a header on each frame which announces that the owner of particular slots will cease transmission in the following frame, allowing contention for these frames.

Another disadvantage of this protocol is that the number of users must be known in advance to allow slots to be allocated to them.

An alternative implicit reservation system (Crowther et al., 1973) does not have permanent owners allocated to slots, and it may therefore be used even when the number of users is not known. Users seize slots using normal ALOHA contention, but once a slot is seized the user becomes its owner and can continue sending data so long as it has any left. (Rules to prevent a user 'hogging' a slot are often applied.) If the slot falls idle for a frame contention can start on it, all users again being treated as equals. Clearly this system can be used even when the number of slots is less than the total number of users.

2.6.1.2 *Explicit reservations*

In one such technique (Roberts, 1973) one or more slots within a frame is divided into smaller slots, called reservation slots. Users

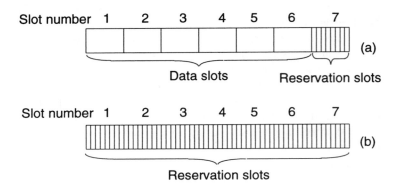

Figure 2.12 Illustration of an explicit reservation ALOHA system: (a) reservation mode; (b) ALOHA mode

contend for these slots, using a technique such as ALOHA, and if successful they place a reservation for a slot over a given number of frames. They can then transmit over these slots without risk of contention from other users. It is therefore required that each user keeps a record of the queue associated with each slot so that they know when to transmit their data. This mode of operation is called the reservation mode.

When there are no remaining reserved slots in the queue (i.e. no stations waiting to transmit data) the whole frame is divided into smaller reservation slots, and it is now in the ALOHA mode. Figure 2.12 illustrates the reservation and ALOHA modes.

2.6.1.3 *Analysis of reservation ALOHA*

The throughput (S_R) of a reservation ALOHA system is affected by whether an end of use flag is included or not in the header of the last packet sent. Such a flag allows users to contend for the slot at the next frame rather than having to wait a frame to discover whether it is empty. Equation 2.20 gives the throughput when no flag is included and 2.21 when it is included. Both equations compare the throughput of the reservation ALOHA with the throughput (S_S) of a slotted ALOHA system.

$$S_R = \frac{S_S}{\frac{1}{k} + S_S} \quad For\ no\ flag \tag{2.20}$$

$$S_R = \frac{S_S}{\frac{1}{k} + S_S - \frac{S_S}{k}} \quad For\ flag \tag{2.21}$$

In these equations k is the average number of packets transmitted before the user gives up his reserved slot. It varies between 1 and infinity.

The maximum throughput of a reservation ALOHA (S_{RM}) is given by Equations 2.22 and 2.23, where the value for slotted ALOHA is given by $\frac{1}{e}$.

$$\frac{1}{1 + e} \leq S_{RM} \leq 1 \quad For\ no\ flag \tag{2.22}$$

$$\frac{1}{e} \leq S_{RM} \leq 1 \quad For\ flag \tag{2.23}$$

2.6.2 Carrier Sense Multiple Access

The Carrier Sense Multiple Access (CSMA) technique relies on the sender sensing the state of the transmission channel and basing its actions on this. It can therefore only be effectively used in channels which have short propagation delays, since for channels with long delays (e.g. satellite) the sensed data is considerably out of date. This is usually specified as having a small α as given by Equation 2.7. In a perfect channel (zero transmission delay) the sender could listen to the state of the channel when it was ready to transmit and only send when the channel was free, so avoiding any collision.

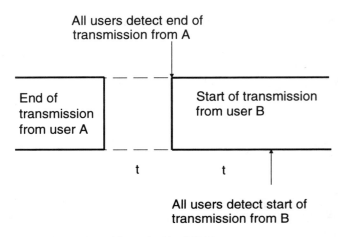

Figure 2.13 Vulnerable period in CSMA

Let t be the time from the start of a transmission from one user to when all users sense the presence of the signal on the line, i.e. the propagation delay on the network plus the sense time. The vulnerable period when collisions can occur is then 2t, i.e. t after the end of one transmission and t to complete a transmission. This is shown in Figure 2.2. If the channels are slotted into t second time slots, then the vulnerable period is reduced to t since transmission must start at the start point of each slot. t is small for low propagation delays on the medium.

CSMA can operate in three modes, 1-persistent, non-persistent and p-persistent. In addition there is a variant called CSMA/CD, as described in the following sections.

2.6.2.1 *1-persistent CSMA*

In the 1-persistent CSMA all users listen to the line prior to transmitting. If they sense traffic they wait. When the line becomes free the users who have been waiting transmit immediately, i.e. transmit with probability one. Collisions will occur if two stations start transmission simultaneously, or if the delay in the transmission channel is such that after one channel has started transmission the second ready channel has not heard this and also starts transmission. After a collision the stations wait a random amount of time before listening to the channel again.

If all transmitted packets are assumed to be of constant length and the load (G) has a Poisson distribution, then the throughput is given by Equation 2.24. It is also assumed in this calculation that all users can sense the transmission of all other users.

$$S = \frac{G e^{-G} (1 + G)}{G + e^{-G}} \qquad (2.24)$$

This equation holds for slotted or unslotted channels, it being assumed that for slotted channels the slot time is much shorter than the packet time.

Figure 2.14 shows the variation of throughput with offered load for different channel delays, including zero delay.

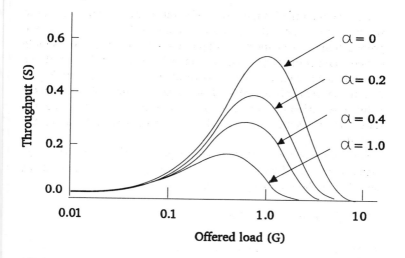

Figure 2.14   Effect of channel propagation delay on the channel throughput for 1-persistent CSMA

2.6.2.2 *Non-persistent CSMA*

In a non-persistent CSMA a user who is ready to send data senses the line, and if it is free it commences transmission. Provided the delay in the transmission line is small (given as a low α) then there is good chance of success.

If the line is busy then the user does not continue sensing, but backs off for a random time before sensing it again, and so on. This method therefore gives a better utilisation of the line, but also results in larger delays than the 1-persistent CSMA.

2.6.2.3 *p-persistent CSMA*

This system is usually used with slotted channels. When a user has data to transmit it senses the channel and if it is free it transmits with probability p i.e. delays transmission to the next slot with probability 1−p.

If the next slot is idle then the user will again transmit with probability p, and so on. If the slot is not free at any time then the user waits a random amount of time and starts again. If initially the user had found that the slot was busy then it would have waited until the next slot and applied the algorithm.

The throughput of the system is given by Equation 2.25, where K is given by Equation 2.26.

$$S = \frac{G\,e^{-G}\,(\,1 + p\,G\,K\,)}{G + e^{-G}} \tag{2.25}$$

$$K = \sum_{k=0}^{\infty} \frac{(\,q\,G\,)^k}{(\,1 - q^{k+1}\,)\,k\,!} \tag{2.26}$$

Figure 2.15 shows the throughput for various CSMA systems and compares this with slotted ALOHA. The delay curves are given in Figure 2.16. Pure contention protocols are best at low loads and have high channel utilisation and low delay. Contention minimisation

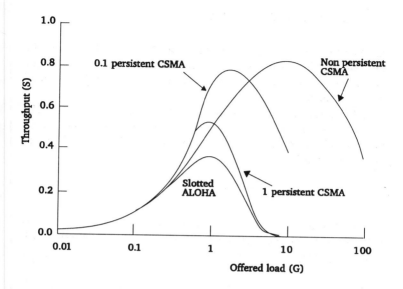

Figure 2.15 Comparison of channel throughput for ALOHA and CSMA, with channel delay factor $\alpha = 0.01$

systems are best at high loads, when they have higher efficiency, but also higher delay at low loads.

2.6.2.4 *CSMA/CD*

In the previous CSMA systems once transmission commences it is completed even if a collision occurs on the first bit. This is clearly wasteful of bandwidth. In a modification, known as Carrier Sense Multiple Access with Collision Detection (CSMA/CD), a user monitors the line even when it commences transmission and stops once a collision is detected (Metcalfe and Boggs, 1976). After that the user waits a random time before sensing the line again. This multiple access method has been standardised by the IEEE as 802.3 (IEEE, 1985).

It is also not always possible for a sender to effectively sense the line during its transmission since the strength of the transmitted signal

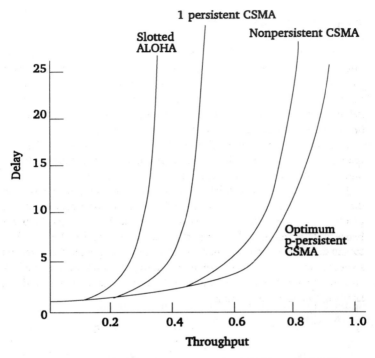

Figure 2.16 Delay-throughput curves for CSMA and ALOHA

may be so strong as to swamp any signal returned back. Often CSMA/CD algorithms require each user that detects a collision transmits a short jamming signal to immediately inform all other users that a collision has occurred on the line.

2.6.3 Asymmetrical protocols

We have so far been considering the case where all users have an equal probability (p) for transmission. This is know as a symmetrical protocol. If somehow it were possible to divide the users into a smaller number of groups, even if this meant splitting the overall channel capacity, then it can be shown that there is a greater overall chance of users being able to transmit without contention, resulting in enhanced throughput.

The ideal system would divide all users in such a way that in any one group there is only one active user, so that the chance of a successful transmission is assured. The trick is in determining a method for allocating users to the available slots within a transmission system. Dynamic allocation is usually used, where the number of users per slot decreases as the load increases.

A dynamic allocation method which has been described (Capetanakis, 1979a and 1979b) is known as the adaptive tree walk protocol, and is illustrated in Figure 2.17 for eight users A to H.

Initially all users who wish to transmit data contend for a slot, and the protocol is at Level 1 and node 1. If a collision occurs then the protocol moves up the tree to Level 2 and node 2. Now only the users below this node (A to D) are permitted to contend for the slot. If a transmission is successful then the slot in the next packet is reserved for users under node 3. If, however, there was contention at node 2 then the protocol moves further up the tree to Level 3 and node 4, limiting the users to A and B.

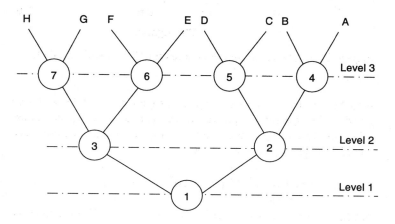

Figure 2.17 Operation of an adaptive tree walk protocol

By moving up the tree each time a collision occurs, the protocol eventually finds users who have information to transmit, and limits their number so that transmission is successful. To ensure fairness users on either side of the tree are searched in turn following a successful transmission or if there is no data available for transmission on one side.

If the loading on the channel is heavy, i.e. many users waiting to transmit, then clearly it is more efficient and quicker to start searching at an intermediate level rather than at Level 1. It can be shown that this optimum level is related to the number of users n who have data to transmit and is given by Equation 2.27.

$$Optimum\ search\ level\ =\ \log_e n \qquad (2.27)$$

Other asymmetrical protocols have been developed, but are not considered here. For example Kleinrock and Yemini (1978) describe the urn protocol. All these have the same aim, to limit the number of users per slot so that the probability of transmission increases, ideally reaching one, with one ready user per slot.

2.6.4 Contention rings

Ring architectures are very powerful and often used in networks. The property of a ring is that the communication line passes through the nodes, and these are not just interfacing to it. The information passing through a node is stored a bit at a time and then forwarded. There is therefore a one bit delay, which can become significant if there are many users on the ring.

Contention rings use a token, similar to the token passing method described later. A token is a unique combination of bits which gives a user authorisation to transmit its data. There is only one token on the line at any time and the only user who can transmit data is the one holding the token, so contention is avoided.

Unlike the token passing method, a contention ring does not have a continuously circulating token, so when none of the users has data to transmit the ring is empty. If a user wishes to transmit it checks to see if a bit is passing though it. If not it commences transmission and at the end of this it puts a token onto the ring. When the token again comes round to the originator he removes it from the ring.

If there is traffic on the ring when a user wishes to commence transmission it waits until the token passes through it. This is then converted to a 'connector', usually by changing the polarity of the last bit, and then it commences transmission. Since there is now no token on the line no other station can transmit.

If two stations start transmission simultaneously, because they both checked the line and did not find a bit passing them, then they will read each other's data as it goes around the ring and will know that a contention has occurred. Both then back off for a random time and start again.

When the ring is full, i.e. many users have data to transmit, the system operates basically as a token passing ring, each user transmitting in turn. On light loads it operates as a contention ring, with low delay since a ready station does not need to wait for a token to arrive before it can begin its transmission.

2.7 Non-contention systems

The previous two sections have examined systems in which pure contention is used in order to access a communication channel and where contention is used but steps are taken to minimise it wherever possible. The present section will look at the large class of systems which avoid contention altogether.

These systems are subdivided into fixed assignment systems, where the channel capacity is divided out amongst users and is fixed; demand assignment systems, where the capacity is divided into segments, but is allocated to users according to their needs at any time; and reservation systems, where the capacity is not rigidly divided and can be reserved in advance by potential users.

2.7.1 Fixed assignment systems

Three techniques are covered under fixed assignment systems: frequency division multiple access (FDMA) which uses the same principles as frequency division multiplexing; time division multiple access (TDMA) which again is based on time division multiplexing principles; and code division multiple access (CDMA) also sometimes referred to as spread spectrum multiple access (SSMA).

2.7.1.1 *Frequency division multiple access*

FDMA works on the principle of dividing the total bandwidth of the communication channel into a number of discrete segments, and allocating each segment exclusively to a user. This is shown pictorially in Figure 2.18. Guard bands are used between each segment of the frequency band to prevent interference between users.

The advantage of the FDMA system is its simplicity, since once the channel capacity is divided amongst users each can operate independently of each other. Since each user has exclusive use of its allocated bandwidth there is no contention and therefore no wastage of bandwidth or delays caused by collisions and retransmissions. Because of this the throughput (S) of non-contention systems equals the offered load (G) if retransmissions due to secondary causes, such as line errors, are ignored. The idealised curve is therefore as in Figure 2.3.

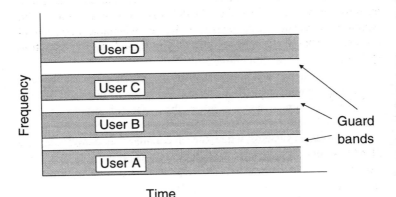

Figure 2.18 User assignment within FDMA

The disadvantage of FDMA systems is that there is wastage of bandwidth, firstly caused by the guard bands and secondly due to the fact that users can only use their own allocated frequency bands. Therefore if a user does not have any information to transmit its allocated band lies idle, even though other users may have a considerable amount of information to send and are experiencing delays on their channel. FDMA is therefore best for use in systems where all users have a stream of data to send, and it is unsuitable for users with 'bursty' traffic, where contention systems, such as ALOHA, perform better.

One of the results of dividing the overall channel capacity amongst multiple users is increased delay (Rubin, 1979). For example, if there are N users sharing a total channel capacity of C and the total traffic they generate is λ with a mean packet length of $1/\mu$, then the delay is given by Equation 2.28.

$$Delay\ D_1 = \frac{1}{\mu\,C - \lambda} \tag{2.28}$$

If the channel is now divided equally amongst the N users, each having exclusive use of its capacity (C/N) only, and the total traffic each user generates is λ/N, then the delay is given by Equation 2.29, which shows the increased delay over the case where the total channel resource is shared amongst all users, using some form of contention system.

$$Delay\ D_2 = \frac{N}{\mu\,C - \lambda} = N\,D_1 \tag{2.29}$$

Another disadvantage of fixed assignment systems, such as FDMA, is that the number of users cannot easily be changed. This would require the overall channel frequency band to be redivided amongst the new users.

FDMA is also not suitable for use in systems which require broadcast of data to many users. Since each user is allocated a single frequency band, it sends on this band and the receiver also monitors this band.

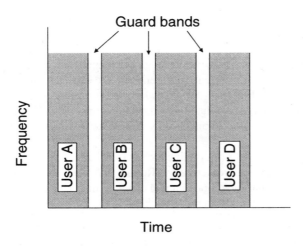

Figure 2.19   User assignment within TDMA

2.7.1.2 *Time division multiple access*

TDMA works on the principle that the complete bandwidth of the channel is allocated to all users, but these users have use of this for a limited time period only (Lam, 1976; Lam, 1977; Hayes, 1984; Stevens, 1990). This is shown pictorially in Figure 2.19.

The user time slots are combined into frames, as in Figure 2.20, which shows a six user system (users A to F). The frames repeat after a frame period of T_F. This is the most usual assignment method where

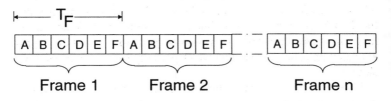

Figure 2.20   The frame structure within TDMA

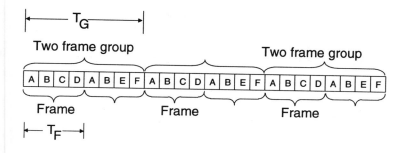

Figure 2.21 Unequal slot assignment within TDMA

each user has equal allocation of the channel. Alternatively some users can have a greater share, and this is accommodated by grouping the frames and varying the user repetition period. This is illustrated in Figure 2.21 where users A and B are allocated increase capacity (with frame repetition period T_F) over users C to F (with frame repetition period T_G).

Because the users always occupy the same position within a frame the receiver knows where to look to collect its data. It is essential to ensure, however, that some form of synchronisation system is used so that user data is correctly timed with the start of their allocated time slots.

As illustrated in Figure 2.19 guard bands (time periods) must be used to prevent interference between users, caused by variations in the synchronisation times. This is shown in Figure 2.22. User A is allocated the first slot and transmits data which is correctly timed but which only occupies a part of its allocated slot. (This again illustrates the inefficiency of fixed assignment systems with fixed packet lengths, when a user does not have data ready to transmit during his allocated time, or only has data which partially fills his allocated time slot.)

Users B and C have data which completely fill their time slots but, due to timing variations, they do not start at the correct times. The guard band period now absorbs this timing variation and prevents interference between adjacent transmissions.

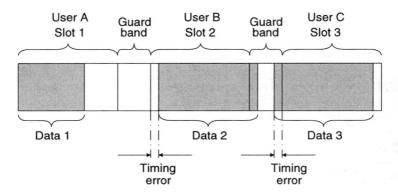

Figure 2.22　Effect of timing variations in TDMA (hypothetical case)

The advantages and disadvantages of TDMA are similar to those for FDMA. It is most suited to systems where the number of users is fixed and each user has a continuous stream of traffic to send. However, since all users are continuously listening to the channel, but only accepting data during their allocated time slots, this method can cope with broadcast systems. Generally TDMA is more efficient than FDMA for the same conditions of operation.

Since there are no collisions and retransmissions within TDMA (as for FDMA) the offered load equals the throughput, if the retransmissions due to transmission errors are ignored. Also, once the time slots have been allocated to users, the performance of any user is unaffected by the loading of other users. Each user experiences a delay determined by the frame period (T_F), the propagation delay within the medium, the transmission time of an average message, and any queueing delays it may have. Assuming that the channel capacity C is equally divided between N users (i.e. each user has full use of the channel for 1/N of the time or has an average capacity of C/N), the normalised transfer delay is given by Equation 2.30.

$$Delay\ D = 1 + \frac{N}{2} + \frac{N\,S}{2\,(1 - S)} \qquad (2.30)$$

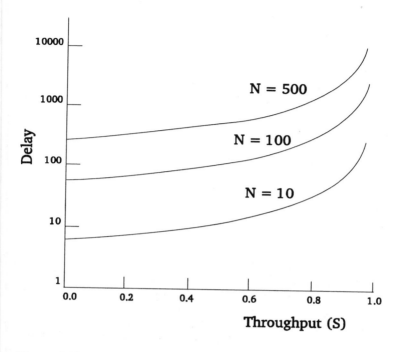

Figure 2.23 Averaged normalised delay within TDMA

Figure 2.23 shows typical average delay curves for TDMA. As expected the delay increases as the throughput increases, due to queuing for access to the user's time slots, and also increases as the number of users is increased, since this results in further division of the communication channel and an increase in the frame repetition period, T_F.

The dependency of delay on number of users is shown more generally in Figure 2.24, so that this is an important parameter in the choice of TDMA systems. These perform best when there is a relatively small number of users, each sending a large amount of traffic. Figure 2.25 shows the reduction of average delay as the average traffic per user is increased.

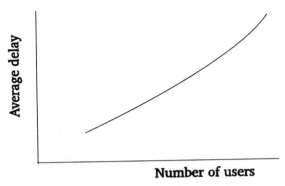

Figure 2.24   Effect of number of users on average delay within a fixed assignment multiple access system

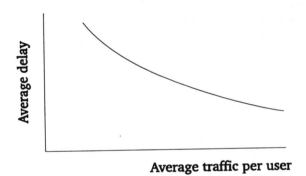

Figure 2.25   Effect of average traffic per user on average delay within a fixed assignment multiple access system

2.7.1.3 *Code division multiple access*

The most common application of CDMA is spread spectrum multiple access (SSMA) and the two terms are often used synonymously (Schilling, 1990; Schilling, 1991).

   CDMA allows the transmission from various users to overlap in frequency and time (unlike FDMA and TDMA, where one of these parameters is used for separation) but segregation between users is obtained by using different codes which are matched between corresponding senders and receivers. (This coding clearly occupies bandwidth.)

SSMA is the most common form of CDMA. Two techniques are used, frequency hopping and phase coding. In phase coded SSMA the phase modulation of the carrier is changed in a coded pattern, whilst in frequency hopping the transmission frequency is varied according to the coded pattern. Figure 2.26 shows a simplified representation of CDMA using frequency hopping. The transmissions are shown synchronised to each other, although in practice no such synchronisation is needed between users. As seen from Figure 2.26 each user 'hops' around between frequency bands, these varying from one time slot to another.

CDMA has several advantages. It has high security against jamming, since no one frequency is used, and for this reason it is used in military systems. In addition it can co-exist with other systems since

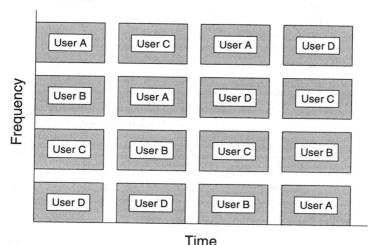

Figure 2.26   Simplified view of user assignment within CDMA (frequency hopping)

separation is via coding, and it is resilient to fading. If fading occurs in a given frequency band then for FDMA all users within this band are affected, but since CDMA only uses a frequency band for a certain portion of time, which varies between users, the effect of fading is spread over many users.

2.7.2 Demand assignment systems

The demand assignment systems are part of non-contention systems in which the communication channel is allocated to a user as demanded, rather than on a fixed basis, as in the previous section.

Four types of demand assignment multiple access techniques are described here: polling, token passing, slotted ring and register insertion.

2.7.2.1 Polling

Polling techniques are ideally suited to star, bus or tree structures although, in practice, any of the systems of Figure 2.1 may be used. The principles of polling have been described and analysed in several texts, such as Betsekas and Gallager (1987), Hammond and O'Reilly (1986), Hayes (1984), Schwartz (1987) and Chou (1983).

In polling each user in turn is interrogated to see if they have any data to transmit. If they have data then they will commence transmission and stop when their buffers are empty. At the end of this they send a 'complete' message and the next user in turn is polled. If a user does not have any data to send he simply returns the 'complete' message. This is therefore a demand assignment system since the more data a user has the longer he is able to transmit. (There may be some rules to prevent a station from hogging all the transmission time.)

Since only one station is polled at any time and is therefore able to transmit, there is no collision. It is also possible to introduce a priority system in which some users are polled more often than others in any period.

The time taken to go around and poll all the users in a system is referred to as its poll cycle. The portion of this period used for overheads (e.g. channel propagation delays, transmission time for the poll messages, synchronisation times when stations receive the poll request) is referred to as the walk time. It is in effect the time needed to transfer permission to transmit from one station to the next, and to complete the messages. Walk time represents the minimum delay time irrespective of whether any data is transmitted.

Three types of polling are considered here: roll call polling, hub polling and probing.

Figure 2.27 illustrates the principle of roll call polling. A central station, which may be a specified user who has taken on this task, polls each of the other users in turn. When a user receives a poll he puts his messages on the common line and the addressed user retrieves it. When all messages have been completed the user sends a 'complete' message to the central station, which then polls the next

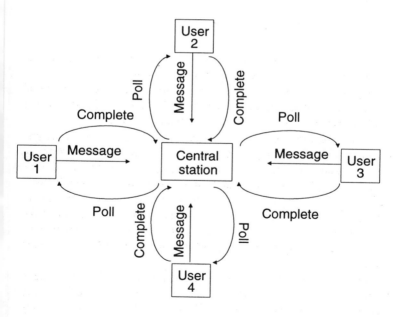

Figure 2.27 Illustration of roll call polling

user in its planned sequence. The central station can be programmed to poll some users more often than others, giving them priority, or it can be made to poll users who consistently transmit large volumes of data more often, so that it acts as a demand assignment system.

In hub polling the central station commences the poll cycle by sending a poll to the first user. If this user has any data to transmit it puts this onto the common line for the addressed user to retrieve. When all data has been completed (of if there was no data in the first place) a 'complete' signal is sent to the next user in line (which need not be the next adjacent user). The 'complete' signal is treated as a poll by the recipient, and so on, as shown in Figure 2.28. Finally the last user in the chain returns the 'complete' signal to the central station, so ending the poll cycle.

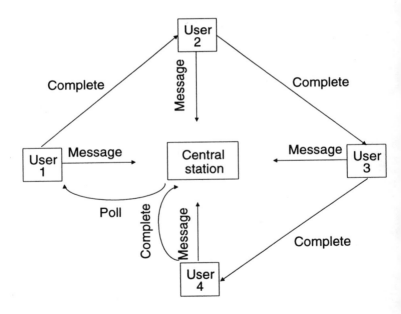

Figure 2.28 Illustration of hub polling

Hub polling can result in reduced walk time between users, since control is not passed back and forth to a central site, although it is more difficult to implement since each user needs greater intelligence.

Assume that there are N identical users, and that all data is held at these user stations in buffers, the buffers being emptied at each poll. The average cycle time (t_c) is related to the average walk time (t_w) by Equation 2.31, where the traffic intensity (ρ) is as in Equation 2.1. Figure 2.29 shows the delay curves.

$$t_c = \frac{N\,t_w}{1 - N\rho} \tag{2.31}$$

The cycle time will therefore be large, leading to poor performance, if the walk time is large, for example as experienced with satellite channels, or if there are a large number of users. This results in large delay.

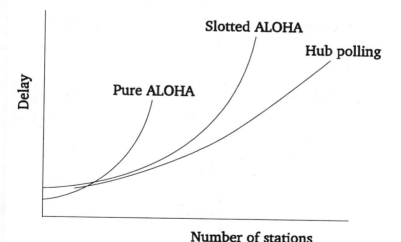

Figure 2.29   Variation of delay in hub polling compared to ALOHA

Polling all users in turn can be a slow process if there are a large number of users and only a few of these have data to send at any time. In these circumstances a technique called probing is used to reduce the number of polls required and so reduce the overall delay in the system.

Probing is a technique in which groups of users are polled at a time, to find the users who are ready for transmission. It is similar to the tree walk protocol illustrated in Figure 2.17. The N users are divided into groups and these are probed simultaneously. The users who have data ready for transmission reply to this probe. For example in Figure 2.17 suppose user F is the only ready user and that node 1 is probed and it provides a ready response. Node 2 is then probed and, as there is no response, the probe is moved to node 3. A response is not received so node 7 is probed (no response) followed by node 6 (response). Finally user E is polled followed by a poll of user F. Therefore, in the worst case, the process of locating and polling the ready user took 5 probes and two polls.

Probing is a good technique to use if the overall loading of the system is light (i.e. few users ready to transmit). If there are 2^n users on the system and only one of them is ready to transmit, then the number of probes (including polls) needed to locate it is given by Equation 2.32.

$$Number\ of\ probes\ =\ 2\,n\ +\ 1 \tag{2.32}$$

If, however, all the users are ready to transmit then the number of probes (and polls) needed is given by Equation 2.33, which is clearly larger than a simple sequential poll method. It is therefore important, during probing, to choose the initial group size so as to minimise the number of probes.

$$Number\ of\ probes\ =\ 2^{n+1}\ -\ 1 \tag{2.33}$$

2.7.2.2 Token passing

Token passing is described and analysed in several papers, such as Bux (1989), Black (1989) and Farber and Larson (1972). It is commonly used in a ring topology and is then referred to as a token ring. It uses active termination points between the users and the network, unlike a polling system which uses passive terminators.

In token passing a special bit pattern, called a token (for example eight logical ones, 11111111) circulates amongst users. This special pattern is prevented from being generated by a user, as part of its normal data, by bit stuffing techniques. Users operate in one of two modes, listen and transmit, as in Figure 2.30. During the listen mode all the bits are read off the line and then regenerated, one bit at a time. If the data is for the user, as indicated by the address, then it is read, otherwise it is just passed on. The user knows a token is present by the special sequence of bits.

Because of the requirement to read and regenerate each bit, there is a bit delay at each user node, which can become quite significant if many users are connected to the line. In addition the delay in the

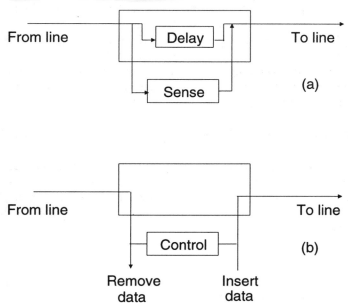

Figure 2.30 Interface to a token passing network: (a) listen mode; (b) transmit mode

overall line must be high enough to permit a whole token to circulate at any time.

If the user wishes to transmit data then it waits for a token to appear. It knows this by the unique bit pattern, and as soon as the token passes through the user's node it is marked as busy. This can be done, for example, by reversing the polarity of the last bit (i.e. 11111110). The user is now in the transmit mode. All the data input from the line is absorbed (since it is not relevant, no user being permitted to transmit without the token) and the data from the user's buffer is emptied out onto the line, immediately after the busy token.

After all the data has been transmitted the user regenerates the token which can now be seized by the next ready user in the chain. When the traffic is light the token circulates from user to user around the ring. When the traffic is heavy the token is seized and marked as busy by each user in turn. As only one token circulates at any time there is no contention.

Token passing is similar to hub polling where no central station is used to commence the poll. Also hub polling is usually applied to bus structures whereas token passing is more often used in ring structures. The scan time in a token passing system is the time it takes for the token to be passed around all users, whilst the walk time is the time taken for a bit to go around the ring (i.e. one bit delay within a user node plus propagation delay in the ring). The performance equations are the same as those derived for hub polling.

In token passing three strategies may be used for the transmitting user regenerating the token. These are:

1. Multiple token operation, where the token is sent straight after the user has completed his data transmission. This allows the next user to 'piggy-back' his data onto it, and is a useful technique to use if the ring is long and has appreciable delays.

2. Single token operation, where the user who has sent data (after having sent the busy token) waits to receive the busy token and his original transmission back again and following the last bit of his data he regenerates the token. This is the system used in the IEEE 802. standards.

3.   Single packet operation, where the token is regenerated after the last bit of the busy token is received back (the user not waiting for his original transmission to be received).

The time for which a token is held by a user can be limited to prevent 'hogging'. If this is long the user will be able to empty his buffers and this is known as exhaustive transmission. If it is very short he will only be able to send a packet every time he receives the token. A master clock is required in all instances to synchronise all users. This clock can be maintained by one of the users (who is referred to as a monitor station). This station also carries out other functions, such as checking for loss of token.

Priority operation is achieved by having a priority slot in the token and assigning each user a priority level. When a user wishes to transmit it waits for a token to pass and, if it is busy, it puts its priority number into the priority slot, if its priority is higher than any which may already be in this slot. The token is then allowed to circulate round the ring. When the original user receives this busy token back it generates a free token and marks it with the priority level which it received. Now only stations with this priority level, or above, are allowed to seize the token.

Figure 2.31 shows typical throughput curves for token passing, comparing this with a contention system such as CSMA/CD. Although the throughput is relatively constant as the number of users increases, above a certain minimum, the delay also increases, as for polling.

IEEE 802.5 describes the token ring protocols, and IEEE 802.4 the token bus. The bus structure can be considered to be a logical ring, as in Figure 2.32, the token moving between users as shown by the dotted line.

2.7.2.3 *Slotted ring*

In the slotted ring (Pierce, 1972) the ring is divided into a number of slots. These slots can hold packets of data, and contain an indication bit which shows whether the slot is empty or full. The ring needs to

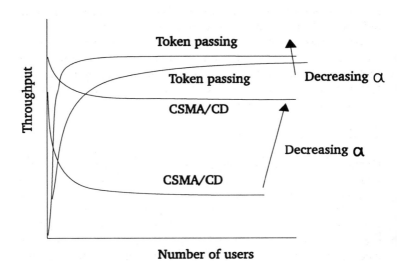

Figure 2.31 Variation of throughput with number of users for token passing and CSMA/CD

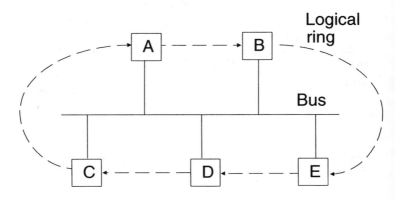

Figure 2.32 Illustration of a logical ring

have a considerable amount of delay, often introduced artificially, in order to hold a reasonable number of circulating slots.

When a user wants to transmit data it waits for an empty slot to arrive. This is then marked as busy and the data in placed in it. The receiver retrieves the data from the slot and once again marks it as free. Clearly the data needs to be the same size as the slot, so this system uses fixed slot sizes.

2.7.2.4 *Register insertion ring*

Operation of this method (Liu, 1978) is best explained by the simplified representation of the interface of a node to the ring, as in Figure 2.33. Data from the line feeds into a shift register and its position is located by a pointer. As the register fills up the pointer moves to the left. Initially the pointer is at the extreme right, indicating that the register is empty. Data is clocked into the register one bit at a time and clocked out again. If the line is very busy the register will be full, but if it is lightly loaded then empty periods will build up and there will be spare room in the register.

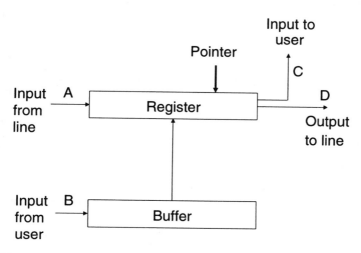

Figure 2.33 A simplified representation of the interface to a register insertion network

When the user wishes to transmit data this is moved into the buffer. It then checks to see there is sufficient room in the register, and if that is the case the data is loaded from the buffer into the register and clocked out onto the line.

If the loading is light the user can continue to send data, but if it is heavy it must wait between transmissions for gaps in the line, and for these to build up in the register. This method therefore automatically prevents users from 'hogging' the line. It also permits variable length packets (as in token passing) and several frames to be on the ring at any time (as in slotted ring).

2.7.3 Reservation systems

There are a large number of different multiple access techniques which are based on some form of reservation. One such system has already been described, that of Reservation ALOHA. This was introduced under contention systems, since some form of contention is used to access the reservation slots, although once this has been done the use of the data slots is contention free.

Reservations can be made using any of the techniques described earlier, pure contention, contention minimisation, or non-contention (fixed assignment or demand assignment). Furthermore, the control mechanism for allocation of slots can be centralised or distributed. In distributed systems each user needs to maintain knowledge of the queue length and its position in the queue. Often all users transmit information of the queue status within their packet headers, so that new users who join the queue can acquire its history. For centrally controlled systems this function would be performed by the master station.

In distributed control it is also required that data and reservation requests from each user are broadcast, so that they are received by all other users, and they know the state of the queue. In a centrally controlled system it is only necessary for the central controller to know about the reservations being made, and for it to be able to instruct the other users.

The efficiency of a reservation system is closely related to the overhead which is used to make reservations. If, for example, the fraction γ of the bandwidth is used for making reservations, as given by Equation 2.34, then the maximum throughput of a reservation channel is given by Equation 2.35.

$$\gamma = \frac{Bandwidth\ occupied\ by\ reservation\ slots}{Total\ channel\ bandwidth} \tag{2.34}$$

$$Maximum\ throughput = 1 - \gamma \quad where\ \gamma < 1 \tag{2.35}$$

Figure 2.34 compares the delay-throughput characteristics of slotted ALOHA and a general reservation systems. Slotted ALOHA reaches a maximum throughput of 1/e (0.368), as in Equation 2.16, whilst typical values of γ for reservation systems are 0.1 to 0.3, giving maximum throughputs from 0.7 to 0.9. However the delay within the reservation systems is higher than in ALOHA, especially when there are a relatively small number of users, and the data is 'bursty'.

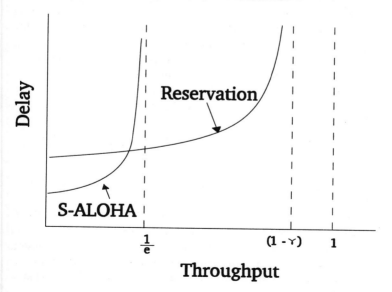

Figure 2.34 Delay-throughput characteristic for slotted ALOHA and reservation systems

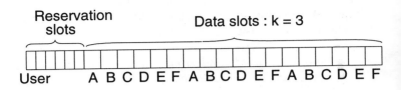

Figure 2.35 Frame structure for R-TDMA

2.7.3.1 *Reservation TDMA*

Reservation TDMA (Weissler et al., 1978) is similar to the reservation ALOHA system except that TDMA is used to access the reservation slots.

The channel is divided into N reservation slots and kN data slots per frame, as in Figure 2.35, where N is the number of users. Each user is allocated one data slot within each of the k subgroups, for his use, and one reservation slot. Reservations are made in the reservation slot and now there is no contention since no other user can transmit in it.

If the owner of a data slot does not wish to use it then it is assigned to other ready users. This is usually done in a round robin fashion.

2.7.3.2 *Bit-map protocol*

This uses a structure similar to R-ALOHA, with explicit reservations, except that no contention is used to access the reservation slots. Figure 2.36 illustrates its operation for six users, A to F.

If there are N users, then initially there will be N reservation slots with each user allocated exclusive use of one of these reservation slots. If a user wishes to transmit it puts a '1' into its allocated slot, and if it does not wish to transmit then it leaves this slot empty. After all the reservation slots have been so accessed, all stations know which users wish to transmit.

Priority is now given to low valued users. The first station who puts a '1' in its slot transmits in the first data slot following the end of the

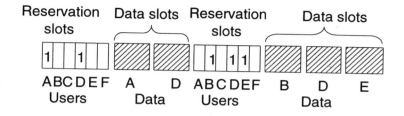

Figure 2.36 Illustration of bit-map protocol

reservation slots. In Figure 2.36 this is user A. The next slot is allocated to the next user in sequence who marked its reservation slot, i.e. user D.

When all the users who reserved slots have transmitted, the reservation slots begin again. If the load is low then reservation slots will repeat continuously until a user is ready to transmit and makes a reservation.

If a user is ready to transmit data after its reservation slot has passed it must wait until the next cycle for it to reappear before it can make a reservation and transmit. The worst case delay occurs for a user located towards the end (user F in Figure 2.36). On average it must wait N/2 reservation slots before it has access to its reservation slot and then, if all users have made reservations, it must wait a further (N-1) data slots before it can begin transmission. To this must be added the transmission propagation time and access times, to obtain the total delay. Users who are allocated reservation slots at the beginning of the reservation cycle (user A in Figure 2.36) only have an average delay of N/2 plus propagation and access delays.

2.7.3.3 *The BRAP and MSAP*

The Broadcast Recognition with Alternating Priorities (Chlamtac, 1976) and the Mini Slotted Alternating Priorities (Scholl, 1976) protocols are very similar in operation, and overcome the two major limitations of the bit-map protocol. These are:

1.   The delay under light load, caused by ready users who have made a reservation but have to wait until the end of the reservation cycle before they can transmit.

2.   The asymmetry in the protocol, where some users get preferential treatment over others, depending on their position within the reservation frame.

Figure 2.37 illustrates the operation of the MSAP and BRAP protocols. As for the bit-map protocol reservation slots are allocated for exclusive use of a user. However as soon as a user makes a reservation it completes its transmission in a data slot. Also the reservation slot sequence now does not start from the beginning but from the last user who transmitted. These therefore overcome the two limitations of the bit-map protocol given earlier. However every user still has to wait for a reservation slot, and has an average delay of N/2 reservation slots.

In a variant of the MSAP a user who has just transmitted is given the option to transmit immediately in another data slot, and only if it does not do this does control pass to the next user. The system therefore gives improved performance under conditions of 'stream' traffic, although it does result in an idle slot under conditions of light traffic, if the user does not take up the invitation to transmit again.

MASP and BRAP may be considered to be similar in operation to roll call polling, but they have a shorter walk time. Generally they have better performance than polling (Kleinrock and Scholl, 1977) but are more complex since users need to be intelligent to make and recognise reservations, and be synchronised to the reservation slots.

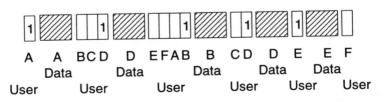

Figure 2.37   Principle of the MSAP/BRAP protocols

2.7.3.4 *MLMA*

The Multi-level Multi-Access protocol (Rothauser and Wild, 1977) overcomes the problem of high delay on light loads, experienced by MSAP/BRAP when there are many users. For example for a 1000 user system the worst case delay could be 1000 reservation time slots with a mean of 500 slots.

In MLMA a user broadcasts its reservation in a given format, for example radix 10, as illustrated in Figure 2.38 for 100 users. Suppose that users numbered 45, 49, 66 and 61 wish to transmit data. They will first broadcast their most significant station number in Frame A. The system now knows that any number of twenty users in the 40s and 60s group wish to transmit.

In Frame C the 40s users are invited to place their reservations, indicating their second digit. Stations 45 and 49 do so and are therefore positively identified. Similarly in frame D users 61 and 66 identify themselves by their second digit. Therefore in this example 30 reservation time slots are used as opposed to 100 which would be needed in MSAP/BRAP. As before each user keeps track of how many reservations have been made and the queue length.

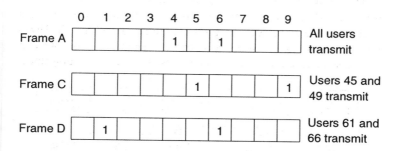

Figure 2.38 Illustration of MLMA

2.7.3.5 *PODA*

The Priority-Oriented Demand Assignment protocol (Jacobs and Lee, 1977; Jacobs, Binder and Hoverstein, 1978; Hsu and Lee, 1978; Jacobs et al., 1979) is a demand assignment system. Like R-ALOHA the channel is divided into frames and each frame has data and reservation slots. Sophisticated scheduling and control algorithms are used, which makes this protocol unique.

Users make reservations for a number of slots and are allocated a priority for transmission on data slots. (There can be two types of data slots, one reserved for particular users and one shared.) A central station then allocates the data slots to users according to their priority and demand. There are usually three classes of traffic, priority, normal and bulk, and these are served in that order.

Fixed assignment can be used to access the reservation slots, when the protocol is called Fixed assignment PODA (FPODA). Alternatively, when there are a large number of users, with light traffic, contention can be used for access to the reservation slots, the system being called Contention PODA (CPODA). However a central station still manages the allocations based on reservations.

2.8 References

Abramson, N. (1977) The Throughput of Packet Broadcasting Channels, *IEEE Trans. Commun.*, **COM-25**, January.

Betsekas, D. and Gallager, R. (1987) *Data Networks*, Prentice-Hall.

Binder, R. et al. (1975a) ALOHA Packet Broadcasting — A Retrospect, *Proc. NCC*.

Binder (1975b) A Dynamic Packet Switching System for Satellite Broadcast Channels, *Proc. ICC*, pp. 41-1 to 41-5a.

Black, U. (1989) *Data Networks: Concepts, Theory and Practice*, Prentice-Hall.

Boothroyd, D. (1995) Spread spectrum solves problems, *New Electronics*, 25 April.

Bux, W. (1989) Token Ring Local-Area Networks and Their Performance, *Proceedings of the IEEE 77*, (2), February.

Capetanakis, J.I. (1979a) Generalised TDMA: The Multi-Accessing Tree Protocol, *IEEE Trans. Commun.*, **COM- 27**, pp. 1476–1484, October.

Capetanakis, J.I. (1979b) Tree Algorithms for Packet Broadcast Channels, *IEEE Trans. Inf. Theory*, **IT-25**, pp. 505–515, September.

Chen, M.S. et al. (1990) A Multi-Access Protocol for Packet-Switched Wavelength Division Multiaccess Metropolitan Area Networks, *IEEE J. Selected Areas in Commun.*, **8**, (6), August.

Chlamtac, I. (1976) *Radio Packet Broadcasted Computer Network — The Broadcast Recognition Access Method*, MS thesis, Tel Aviv University.

Chou, W. (ed.) (1983) *Computer Communications*, Prentice-Hall.

Clare, L.P. et al. (1990) A performance comparison of control policies for slotted ALOHA frequency-hopped multiple access systems. In *Proceedings of the new era IEEE military communications conference, MILCOM 90, (2), September.*

Clark, D.D., Pogran, K.T. and Reed, D.P. (1978) An Introduction to Local Area Networks, *Proc. IEEE*, **66**, pp. 1497–1517, November.

Crowther, W. et al. (1973) A System for Broadcast Communication: Reservation ALOHA, *Proc. 6th Hawaii Int. Conf. Syst. Sci.*, pp. 371–374.

Dono, N.R. (1990) A Wavelength Division Multiple Access Network for Computer Communication, *IEEE J. on Selected Areas in Commun.*, **8**, (6), August.

Farber, D.J. and Larson, K.C. (1972) The System Architecture of the Distributed Computer System — the Communications System, *Symp. on Comput. Networks*, Polytechnic Institute of Brooklyn, April.

Gerla, M. and Kleinrock, L. (1977) Closed Loop Stability Controls for S-ALOHA Satellite Communications, *Proc. 5th Data Commun. Symp.*, pp. 2-10 to 2-19.

Hammond, J.L. and O'Reilly, P.J.P. (1986) *Performance Analysis of Local Computer Networks*, Addison-Wesley.

Hayes, J.F. (1978) An Adaptive Technique for Local Distribution, *IEEE Trans. Commun.*, **COM-26**, August.

Hayes, J.F. (1984) *Modelling and Analysis of Computer Communication Networks*, Plenum Press.

Hsu, N. and Lee, N. (1978) Channel Scheduling Synchronisation for the PODA protocol, *Proc. Internat. Conf. on Commun.*, Toronto, June.

Hwa, H.R. (1975) A framed ALOHA system, *Proc. PACNET Symposium*, Sendai, Japan, August.

IEEE (1977) Special Issue on Spread Spectrum Communications, *IEEE Trans. Commun.*, **COM-25**, August.

IEEE (1985) *IEEE 802.3: IEEE Standards for Local Area Networks — Carrier Sense Multiple Access with Collision Detection (CSMA/CD) Access Method and Physical Layer Specifications*, New York, IEEE.

Jacobs, I.M. and Lee, L.N. (1977) A Priority-Oriented Demand Assignment (PODA) protocol and an Error Recovery Algorithm for Distributively Controlled Packet Satellite Communication Network, *EASCON '77 Convention Record*, pp. 14-1A to 14-1F.

Jacobs, I.M., Binder, R. and Hoversten, E.V. (1978) General Purpose Packet Satellite Networks, *Proc. IEEE*, **66**, (11), November.

Jacobs, I.M. et al. (1979) Packet Satellite Network Design Issues, *Proc. National Telecom. Conf.*, November, pp. 45.2.1–45.2.12.

Ketseoglou, T. and Zimmerman, T. (1994) A likely candidate, *Communications International*, March.

Khansefid, F. et al. (1990) Performance analysis of code division multiple access techniques in fibre optics with on- off and PPM pulsed signalling. In *Proceedings of the new era IEEE military communications conference, MILCOM 90, (3) September.*

Kleinrock, L. and Lam, S. (1973) Packet switching in a slotted satellite channel, *Nat. Computer Conf., AFIPS Conf. Proc.*, **42**, AFIPS Press, pp. 703–710.

Kleinrock, L. and Tobagi, F. (1975) Random Access Techniques for Data Transmission over Packet-Switched Radio Channels, *Proc. NCC*, pp. 187–201.

Kleinrock, L. (1976) *Queueing Systems*, John Wiley.

Kleinrock, L. and Scholl, M. (1977) Packet Switching in Radio Channels: New Conflict-Free Multiple Access Schemes for a

Small Number of Data Users, *Conf. Rec. Int. Conf. Commun.*, Chicago, June.

Kleinrock, L. and Yemini, Y. (1978) An Optimal Adaptive Scheme for Multiple Access Broadcast Communications, *Proc. ICC*, pp. 7.2.1 to 7.2.5.

Ko, C.C., Lye, K.M. and Wong, W.C. (1990) Simple priority scheme for multichannel CSMA/CD local area networks, *IEE Proc.,* **137**, (6), December.

Lam, S.S. and Kleinrock, L. (1975) Packet Switching in Multiaccess Broadcast Channels: Dynamic Control Procedures, *IEEE Trans. Commun.*, **COM-23**, pp. 891–904, September.

Lam, S.S. (1976) Delay Analysis of a Packet-switched TDMA System, *Proceedings of the National Telecommunications Conference*, December.

Lam, S.S. (1977) Delay Analysis of a Time Division Multiple Access (TDMA) Channel, *IEEE Transactions on Communications*, **COM-25**, (12), December.

Liu, M.T. (1978) Distributed Loop Computer Networks. In *Advances in Computers*, M.C. Yovits (ed.), Academic Press.

Mark, J.W. (1978) Global Scheduling Approach to Conflict-Free Multiaccess via a Data Bus, *IEEE Trans. Commun.*, September.

Metcalfe, R.M. and Boggs, D.R. (1976) Ethernet: Distributed Packet Switching for Local Computer Networks, *Commun. ACM*, **19**, pp. 395–404, July.

Morris, M.J. and Le-Ngoc, T. (1991) Rural telecommunications and ISDN using point-to-multipoint TDMA radio systems, *Telecom. J.*, **58**.

Pierce, J. (1972) How Far Can Data Loops Go? *IEEE Trans. Commun.*, **COM-20**, pp. 527–530, June.

Roberts, L. (1972) Extensions of Packet Communications Technology to a Hand Held Personal Terminal, *Proc. SJCC*, pp. 295–298.

Roberts, L. (1973) Dynamic Allocation of Satellite Capacity through Packet Reservation, *Proc. NCC*, pp. 711–716.

Roberts, L. (1975) ALOHA packet system with and without slots and capture, *Comput. Commun. Rev.*, **5**, pp. 28–42, April.

Rodrigues, M.A. (1990) Evaluating Performance of High Speed Multiaccess Networks, *IEEE Network Magazine*, May.

Rothauser, E.H. and Wild, D. (1977)MLMA — A Collision-Free Multi-Access Method, *Proc. IFIP Congr. 77*, pp. 431–436.

Roy, R. and Warwick, M. (1994) Journey into space, *Communications International*, August.

Rubin, I. (1979) Message delays in FDMA and TDMA communication channels, *IEEE Trans. Commun.*, **COM-27**, May.

Schilling, D.L. et al. (1990) CDMA for personal communications networks. In *Proceedings of the new era IEEE military communications conference, MILCOM 90, (2), September.*

Schilling, D.L. et al. (1991) Spread Spectrum for Commercial Communications, *IEEE Commun. Magazine*, April.

Scholl, M. (1976) *Multiplexing Techniques for Data Transmission over Packet Switched Radio Systems*, PhD thesis, UCLA.

Schwartz, M. (1987) *Telecommunication Networks: Protocols, Modelling and Analysis*, Addison-Wesley.

Stallings, W. (1988) *Data and Computer Communications*, Macmillan.

Stevens, D.S. et al. (1990) Evaluation of slot allocation strategies for TDMA protocols in packet radio networks. In *Proceedings of the new era IEEE military communications conference, MILCOM 90, (2), September.*

Strole, N.C. (1987) The IBM Token-Ring Network — A Functional Overview, *IEEE Network*, (1), January.

Tsiligirides, T. and Smith, D.G. (1991) Analysis of a p-persistent CSMA packetized cellular network with capture phenomena, *Computer Communications*, **14**, (2), March.

Weissler, R.R. et al. (1978) Synchronisation and Multiple Access Protocols in the Initial Satellite IMP, *Fall COMPCON*, September.

Yoon, J. and Baffer, B. (1993) Multiple access technology boosts mobile radio capacity, *Cellular & Mobile International*, September/October.

3. Packet switched networks

3.1 History and underlying concepts

Packet switching is a mature, secure method of transferring data across a network. It divides data into segments, each of which is wrapped in an envelope to form a packet; a typical message comprises one or more packets. Each packet contains the actual user data plus information helpful to its movement across the network, such as addressing, sequencing and error control.

A tried and tested, robust technique, packet switching attracts a large and growing installed base. It is ideal for transaction-orientated applications such as banking, point-of-sale retail and electronic mail. Investment in packet switching continues to grow at over 25% per annum, largely in the private sector, while most industrialised countries offer at least one public data network based on the technology.

The driving force behind the development of packet switching was the need in the late 1960s for asynchronous terminals to access numerous remote hosts. Its rapid acceptance owes much to the wide implementation of the X.25 international standard interface to packet switched networks (PSNs).

While private leased lines offer fixed bandwidth, security and protocol transparency, data transfer speed is limited by the capability of the attached modems, re-routeing is impossible, unused bandwidth is unavailable to other users and a charge is levied for the line irrespective of use.

Circuit switching, as found in a conventional telephone network, provides a dedicated circuit for the duration of the call and charges according to time and distance. However, public voice networks are optimised for voice, prone to data-corrupting noise and are normally without security, such as call screening, for dial-up access to a host computer.

As networks have grown in size and importance, functions such as routeing, error correction and flow control have also become essential.

Packet switching therefore evolved to distribute these components throughout a network. It is a subset of traditional message switching, in which data is transmitted in blocks, stored by the first switching node it meets in the network and forwarded to the next and subsequent downstream nodes until it reaches the destination.

At its heart is the concept of the virtual circuit: a fixed path through the network from sender to destination is defined at the beginning of the session or call. The path remains unchanged for the duration of the connection. (Figure 3.1).

The idea owes much to statistical multiplexing, in which many switched circuits can be active on a single physical link – an efficient use of a circuit's available bandwidth. Extending the idea beyond statistical multiplexing, however, packet switching offers the following advantages:

1. Each terminal in a group sharing the same physical circuit may be connected to a totally different destination. This versatility is one of the major strengths of packet switching. (Figure 3.2).

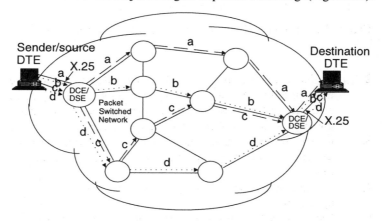

Figure 3.1 The virtual concept: multiple paths

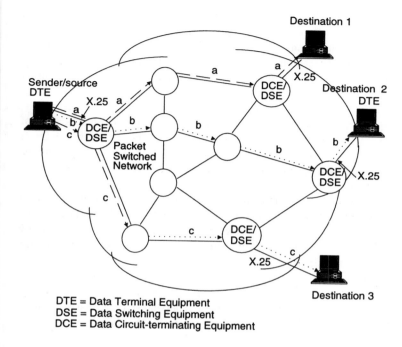

DTE = Data Terminal Equipment
DSE = Data Switching Equipment
DCE = Data Circuit-terminating Equipment

Figure 3.2 The virtual circuit concept: multiple destinations

2. Unlike message switching, packet switching has a block size limit, usually 128 bytes or 256 bytes, but up to 4096 bytes on some networks, thereby reducing storage demands or buffering in the nodes, as well as delay at each node.

3. No single user or large data block can tie up circuit or node resources indefinitely, making it well suited for interactive traffic.

4. Data protection against corruption or loss; errors are corrected by retransmission.

5. Users can select different destinations for each virtual call, overcoming the inflexibility of point-to-point dedicated networks.

6.   Simultaneous calls allow PC users to access multiple windows to different remote applications.
7.   Since many users can share transmission resources efficiently, the cost of intermittent data communications is reduced.
8.   New calls can be added and old ones disconnected without affecting other users.

While suppliers have increased the number and sophistication of switch functions over the years, supporting a myriad of terminal and host protocols, the X.25 specification of an interface to PSNs remains standard and non-proprietary, offering users flexibility in equipment choice.

The first recommendations for a standard PSN interface were laid down in 1976 by a subcommittee of the ITU-T (formerly CCITT). They have evolved into the X.25 international standard, approved by the International Standards Organisation in 1984 (ISO 8208), for accessing a host over a wide area network.

Geographically scattered standalone LANs can be interconnected to form a single logical LAN using X.25 over WAN links, taking advantage of its shareable, protocol independent data transport mechanism. Built under the Open Systems Interconnection framework, the network can be upgraded to take advantage of new developments, leaving the software affecting the upper layers in place.

Significantly, the recommendations say nothing about the internal protocols and algorithms needed to run a PSN but describe the interface between Data Terminal Equipment (DTE) and Data Circuit-terminating Equipment (DCE) for devices designed to run in packet mode. Rather than being a protocol in itself, the X.25 specification describes an interface using individual protocols at the three lowest layers of the OSI model.

3.2 Packet switched call

Each call on a PSN is made up of several steps (Figure 3.3), similar in their progression to a switched telephone call. A terminal user enters a command for connection to the network, then the numeric network address of the destination.

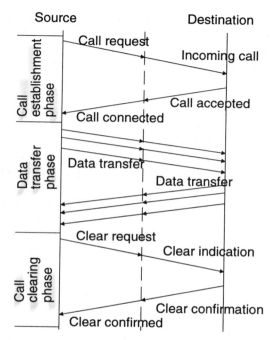

Figure 3.3 Phases of a PSN call

On hitting the Enter key, a specially formatted call request packet is generated and sent across the network; it holds the network address of both the source and destination. If the call is unacceptable, the destination host may reject the call and disconnect. If the call is accepted, a call connected packet is returned to the sender, indicating that an end-to-end switched virtual circuit (SVC) has been established for the duration of the call.

To either end, the terminal and host, the SVC looks like a circuit switched connection, but because packets can take any route in the network, call set-up time is less, often measured in tenths of a second compared to several seconds.

Typically, the first PSE in the chain determines to which particular link the call request packet will be forwarded and sends it to the next

PSE along that route. This continues until the data reaches its destination. Once connection is made, the data transfer begins: electronic mail, updates to databases and so on. It is the network's responsibility to ensure fast, error-free data delivery, transparently between the two end points.

Network operations at this stage include sorting the packets into the correct sequence at the destination, flow control (ensuring data input rates match delivery rates) and signalling, i.e. notifying the source or destination of any unusual network events.

The final phase of a packet call is disconnection or call clearing, by either the source or destination. A unique packet notifies the network and the opposite end that the call is to be cleared and the SVC freed. Each user-originated call over the network demands its own virtual circuit, whether a SVC, emulating the PSTN, or permanent link, such as a leased line. For a permanent circuit, the DTEs are permanently connected, obviating the need for a call set-up phase.

3.3 The packet switched network

A DTE is a generic term for any equipment designed to manipulate packets that is attached to the network (Figure 3.4). It could be an end-user terminal or a standalone minicomputer running X.25 software and offering an X.25 port, for example. It could also be a Packet Assembler/Disassembler (PAD), converting data between X.25 and other communications protocols, allowing devices unable to support X.25, such as ASCII terminals, to access an X.25 based PSN.

The term PAD is often applied too widely, covering any device allowing non-X.25 devices to communicate over an X.25 based network, such as devices supporting SNA to X.25 conversion.

A PAD assembles packets as data arrives from attached users in asynchronous, character by character, form. It assigns relevant addresses and error correction data and forwards the packets on to the switch. Conversely, it sorts data arriving from the switch in packets, converting the packets into a character stream again for onward passage to the recipient computer.

PADs are most widely used for asynchronous display terminals but can support printers, terminals servers and other devices. There need

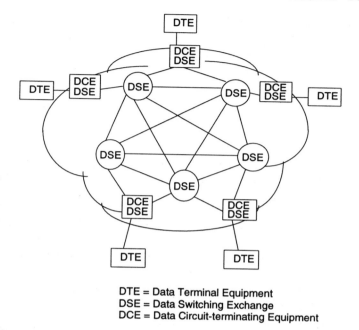

DTE = Data Terminal Equipment
DSE = Data Switching Exchange
DCE = Data Circuit-terminating Equipment

Figure 3.4 PSN mesh topology

not be a PAD at both ends of the call – one end may be a PAD tackling a terminal's asynchronous signals, the other might be a mainframe host with a direct connection through an X.25 port.

PADs invariably support three ITU-T standards: X.3, X.28 and X.29, allowing each PAD port to match the requirements of incoming terminals and remote host computers. X.3 defines PAD parameters, such as whether or not it echoes keystrokes back to the screen; X.28 and X.29 control how those parameters are installed, X.28 from terminal to PAD via a keyboard command and X.29 from host to PAD via a so-called qualifier bit in a data packet.

Each X.25 port in the network and all the devices associated with it such as modems and cables, constitute a DCE.

No ITU-T specification exists for DCE-to-DCE communications, so data must pass through an intermediate device known as a Data or

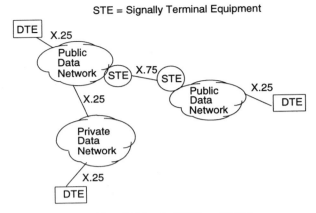

Figure 3.5 Interworking PSNs via X.25 and X.75

Packet Switching Exchange (PSE), either at an intermediate or at a destination point on the network. PSEs are usually linked in a mesh topology, providing at least two circuits to each node, their bandwidth determined by traffic demand. (Figure 3.5).

While the ITU-T standards offer guidance for access to a PSN, what goes on inside it is left entirely to the network suppliers and PSE manufacturers such as Northern Telecom. The internal protocol usually resembles X.25 but adds functionality such as flow control, network management and accounting.

Internetworking between discrete PSNs is increasingly important, such as between private and public networks. It takes place via virtual circuits through gateways. The ITU-T X.75 recommendation supports X.25 to provide such a gateway in special nodes known as signalling terminal equipment (STE). They provide connectivity services but operate at the first three OSI layers in contrast to a full OSI gateway that employs all seven layers.(Figure 3.6).

3.4 Packets and the OSI model

At the Physical, Data Link and Network layers of the OSI seven layer model, equipment at customer premises or in the network must share

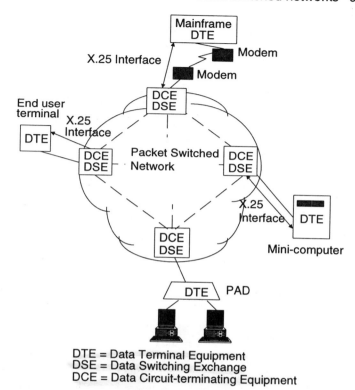

DTE = Data Terminal Equipment
DSE = Data Switching Exchange
DCE = Data Circuit-terminating Equipment

Figure 3.6 Inside and outside the X.25 cloud

a compatible set of interfaces and signalling conventions to comply with the X.25 standard.

Each level or layer is responsible for one aspect of the total X.25 based communication. It is functionally independent from the other layers, interacting with them in clearly defined areas, providing a service to the level above and receiving a service from any lower level. Corresponding levels on the other side of the link, a level's peer, communicate via a peer protocol, co-ordinating the exchange of information between peer levels across the interface. (Figure 3.7).

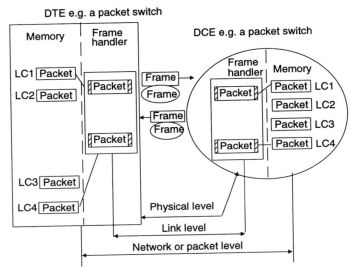

Figure 3.7 X.25 packet switch recommendations for PSN access

The higher layers proposed by the ISO, through which data passes before reaching the packet switching layers, have not been fully laid out, though their overall aims are well defined. Once at the X.25 level, communication can be established between an end point device and the corresponding X.25 level of the local DCE. Having passed the local DTE/DCE interface, a conversion to the network protocol takes place for transport through intermediate nodes in the network. (Figure 3.8).

At the destination DCE, conversion back to the X.25 protocol takes place for transporting the message across the DCE/DTE interface, the data then moving up each of the seven layers until it reaches the final application at the destination end point device.

3.4.1 Packet switching at the Physical Layer

The Physical Layer describes the physical connection — the electrical interface and procedures involved in establishing a communica-

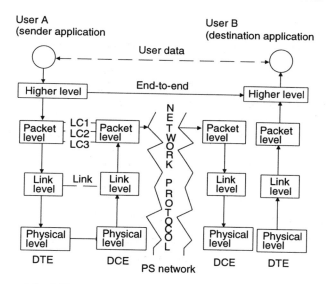

Figure 3.8 PSN transportation through the OSI layers

tion path. The connection outside the X.25 cloud is between an end point device, such as a user terminal, and a PSE on the network. Within the cloud, connections are between the PSEs themselves as they make up the network. Each network link will have different characteristics, chosen according to many parameters such as the maximum acceptable error rate, minimum rate of transfer and the distance over which the circuit must run.

The most common medium for DTE to DCE connection is copper via a leased line to a public data network, but fibre links are becoming increasingly cost effective. Depending on the electrical signalling, devices may be connected directly, otherwise line drivers or modems are necessary to carry the signals over long distances.

Up to about one million bits per second over distances below about one kilometre, twisted pair (ordinary telephone wire) is normally adequate. Beyond this fibre optic cable, which is insensitive to electrical noise, may be required to ensure the error rate on the link is not so high that the error detection and correction at layer 3 is rendered ineffective.

The ITU-T chose the X.21 specification as the physical interface between a DTE and DCE for X.25, despite the prevalence of the Electrical Industries Association's RS-232-C standard for computer-modem connections. There are only minor differences, however, with X.21 geared eventually to delivering a digital signal to customer premises, replacing the analogue circuit switched network with ISDN.

Prior to ISDN, the ITU-T approved an interim standard, X.21bis, which is practically identical to the ubiquitous RS-232-C. It specifies a leased line connection, but dial-up is feasible.Information passing across the interface at this level comprises three signal types:

1. Data signals transfer the higher level information between the two devices, a signal carrying data in one direction and another in the other direction.

2. Clock signals, in conjunction with data signals, reconstruct transmitted information at the destination.

3. Where devices are connected directly via an X.25 port, software handles flow control (which monitors the amount of traffic on the circuit) automatically as part of the protocol functions. Asynchronous traffic, however, makes use of control signals to manage the link traffic volumes and to decide whether the circuit is in an acceptable condition.

In-band or software flow control involves special control characters or bit patterns sent by the receiver, such as a terminal or a PAD, which are embedded in the normal data stream to tell the transmitter to slow down its data output.

After a receiving device has submitted a clear-to-send signal, it has the option of dropping the command at any time if traffic becomes too heavy, a process known as out-of-band or hardware flow control. This method in effect stretches the function of an RS-232 command originally intended for modem control. The whole subject of flow control outside of software based controls, though, is the subject of complex debate.

Within X.25, much of the responsibility for flow control is given to layers 2 and 3. It should be remembered, too, that the Physical Layer takes no account of line transmission quality. It passes valid and corrupted data with equal enthusiasm along the circuit; its job is only to hide the nature of the physical media from the Data Link Layer above.

3.4.2 Packet switching at the Data Link Layer

The aim of the Data Link Layer is to ensure that data passing between devices is error free and in the correct order. Given the global influence of the X.25 standard, it is no surprise that the most important PSN link layer in use is X.25 layer 2, although there are other variations of layer 2 implementation that achieve the same objective.

Link layer information is passed between the two end point devices in frames. The most common protocol used for encapsulating data into frames is the ISO's High-Level Data Link Control (HDLC) dating back to the mid-1970s. Its origins lie in work done by IBM seeking to replace older Binary Synchronous Communications with its Synchronous Data Link Control protocol.

Unlike previous protocols, HDLC, SDLC and their peers, such as ANSI's Advanced Data Communications Control Procedure, are bit oriented rather than character oriented. Frame length in a bit oriented protocol is an arbitrary number of bits rather than a fixed multiple of a selected character size. This makes life simpler by reducing the number of necessary framing and control characters. The ITU-T adopted HDLC largely intact, producing the Link Access Procedure Balanced (LAPB).

3.4.2.1   *The make-up of a frame*

A basic X.25 level 2 or HDLC (Figure 3.9) frame opens and closes with a flag, a specific pattern of bits transmitted on the link to act as reference points for frame synchronisation. When there are no frames to transmit on a circuit, flags are sent continuously to fill in time and maintain synchronisation between the two end points.

Flag 01111110	Address field	Control field	Information field	Frame check sequence	Flag 01111110
8 bits	8 bits	8 bits	N bits	16 bits	8 bits

Figure 3.9 An X.25 level two frame: HDLC structure

When a recipient detects a flag, it examines the data stream for an eight bit address field or area, used to identify to which direction of transfer the frame belongs. A control field specifies what the frame contains and carries frame sequence numbers, acknowledgements, retransmission requests and other control information.

Next comes the information field, containing the actual user data, followed by a frame check sequence (FCS) of bits by which the receiver determines, by reference to a checksum, whether or not frames have been received correctly.

A final or trailing flag marks the end of the frame and indicates to the receiver that the preceding 16 bits should be interpreted as the FCS field. The recipient processes the data stream, compiles its own FCS and compares the result with the FCS received in the frame. If they match, the data is correct, otherwise the frame is discarded and the receiver issues a retransmission request.

HDLC and LAPB contain three types of frame, each built up from the building blocks outlined above:

1. Unnumbered frames carry no send or receive sequence numbers and simply offer additional data link control functions such as circuit initialisation and disconnection, helping to set up and later close down or reset links.
2. Supervisory frames control information flow, request retransmissions and acknowledgement frames. They contain commands such as Receive Ready from a DTE or DCE or a temporary busy condition, for example.
3. Information frames are the only frames to transport data across the link.

To establish the link under LAPB, either end of the DTE/DCE link can begin link initialisation by sending out the relevant command, known as the Set Asynchronous Balanced Mode (SABM). Acceptance of the SABM is confirmed by the other end when it issues an Unnumbered Acknowledgement response. Once this process is complete, I-frames can begin to flow.

3.4.2.2 *Error correction and flow control*

As mentioned above, the Data Link Layer is responsible for error correction and flow control of I-frames. The most common technique is known as positive acknowledgement. A recipient must acknowledge that a frame has been received correctly. Alternatively, but less common, the receiver may only respond to a received frame if it is not received correctly: negative acknowledgement.

At its most basic, the transmission process works as follows. A sender transmits a frame then waits for an acknowledgement from the recipient. If the checksum is correct, the frame is regarded by the receiver as correctly received and the receiver sends back an acknowledgement. If the checksum is incorrect, the frame is regarded as corrupted. The receiver discards that frame and simply waits.

If the sender successfully receives an acknowledgement, it assumes the frame has been correctly transferred and can send the next. If nothing arrives within a set acknowledgement period, it assumes the frame was not correctly received and re-transmits, waiting again for a response. If this is repeated a set number of times, the sender assumes the link is faulty.

There are, however, complications such as frame corruption, an incorrectly received acknowledgement and even the appearance of false frames due to electrical interference. To help identify one frame from another, each information frame is assigned a pair of sequence numbers, carried for X.25 level 2 in the control field, operating in both the send and receive directions.

The initial frame carries a sequence number in a three bit field. The receiver, maintaining a count of error free frames received, sends an acknowledgement carrying the sequence number of that first frame then adds one to the sequence number, which is the anticipated number of the next frame. The sender receives the acknowledgement, adds one to the sequence number, attaches it to the next frame and repeats the process, which is known as a Modulo-8 or Modulo-N frame level window.

If, therefore, a receiver's reply is corrupted, the transmitter can ignore the frame and wait for the acknowledgement time to expire. It then sends the original frame again: as the receiver has successfully received the frame on the sender's first attempt, it will be expecting a frame marked with the next sequence number along. It can conclude that the sender did not receive a correct acknowledgement, discard the repeated frame and resend the original acknowledgement.

Sequence numbers range from zero to seven, avoiding the problem of large numbers taking up valuable frame space. It could work with just two numbers: in the example above, the initial frame could be labelled 0 and the faulty acknowledgement 1. The receiver will be sent the retransmitted frame, still numbered 0 and can see immediately what has happened.

To avoid the sender having to wait for an acknowledgement before transmitting the next frame, thereby limiting throughput due to the time taken to transmit, check and process frames, X.25 level 2 adopts a window size of seven. This means the sender can transmit up to seven frames before an acknowledgement is required, allowing the receiver to be relatively slow and utilising the circuit more efficiently.

As data will normally flow in both directions on a link simultaneously (full duplex operation) different sequence numbers are allotted for each direction. A device will therefore transmit and receive a mixture of information frames and acknowledgement frames.

Some packet switch protocols, including X.25 level 2, can piggyback an I-frame with an acknowledgement frame that acknowledges a I-frame from the other direction. Although often difficult to implement, it reduces the total number of frames transmitted and received.

The Modulo-N or window-N system of sequence numbering enables the Data Link Layer to detect and correct corrupted frames. The limited window size plus the numbering provides flow control.

Should a receiver be unable to process frames as fast as is necessary at any one point in the communication, it simply stops acknowledging frames it is receiving. The sender transmits as many frames

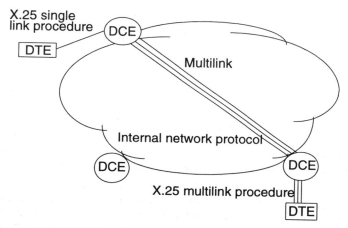

Figure 3.10 Multilink procedure within a PSN

as are allowed before it needs an acknowledgement, then stops and waits, giving the receiver a chance to catch up.

A DTE or DCE can suspend transmission by issuing a disconnect command, acknowledged by the receiver through a UA response as for set-up, after which the link can be closed.

The LAPB protocol also contains a complex variety of commands and responses indicating reasons for retransmission, frame reject responses and so on.

Within the Data Link Layer mandate to ensure error free transmission, the ITU-T X.25 1984 recommendation added a service known as the Multilink Procedure (MLP). The function of MLP is to distribute packets from layer 3 across multiple physical links (Figure 3.10). It is most associated with expensive international rather than national links.

Although packets carried along the same virtual circuit may travel on different physical paths, to the Network Layer, these different physical links appear as one logical circuit. MLP manages any number of Single Link Procedures, each of which is an orthodox LAPB circuit. It carries out resequencing for all received frames over different physical links, adding a multilink control field and an information field to each frame

MLP offers several advantages:

1.   Load sharing or balancing is an inherent strength within MLP.

2.   Improved reliability and resilience against link failure as packets flow via different physical paths.

3.   Additional bandwidth to support increased traffic can be added incrementally without affecting or disrupting existing physical connections.

Packet switch links over satellite are subject to far longer transit delays than normal, around 250ms one way, making X.25 inefficient if used with Modulo-8 windows. The sender will be idle for relatively long periods waiting for acknowledgements.

ISO and the ITU-T therefore produced Modulo-128, within LAPB, allowing up to 127 frames to remain outstanding before an acknowledgement is needed. Currently available only within some public PSNs, it is accessed through an extended SABM command. A future development is likely to be some form of hot keying between Modulo 8 and Modulo 128 modes.

3.4.3 Packet switching at the Network Layer

While the Data Link Layer sorts out an error free link between two connected devices, the Network Layer provides communications between devices that are not necessarily connected together as they will have a network in between them.

Network Layer terminology talks of information in packets rather than in their Data Link equivalent, frames. A Network Layer for an X.25 PSN is connection oriented. A connection is established between two end point devices via a virtual circuit, and making that connection is a major function of this layer.

Much of layer 3's work is transparent to end users, such as multiplexing simultaneous calls over a single physical connection. Routeing, relaying, packet sequencing and flow control are also functions of the Network Layer, plus providing services to the Transport Layer and higher layers, such as addressing and data transfer.

To make a normal telephone call, it is necessary to know the number at the other end. In the same way, Network Layer connections rely on a system of reference that provides a unique address to locate each device attached to the network. Two methods predominate in providing end point addressing for PSN attached devices.

Depending on the size and complexity of the network, each device can be given a unique address that remains valid throughout the network. From anywhere on the network, the device can be contacted via that address and the address remains the same at all times.

Alternatively, the route taken to reach a device is used to build up the address, so that the address differs according to the route taken and, naturally, also depends on the starting point on the network. A device's address is an outline of the route the data will take to reach it.

In Figure 3.11 the network links are numbered from one to nine. An end point device at A sending data to end point device B could take a route along circuit seven to six, six to three, three to one and out to B. From A's viewpoint, B's address would be 7631 along this particular route. For C to send data to B, the shortest route would be via circuit two followed by one. B's address from C's viewpoint would be just 21.

Known as a route addressing system, a feature of this method is that the address of any end point device will vary according to who is sending data, in contrast to the global addressing system in which an end point's address remains the same irrespective of wherever it is being sent data.

A typical call set-up in Figure 3.11 would occur as follows. Device A establishes a virtual call on a virtual circuit to device B by sending a call request packet down its link to packet switch 4. A call request packet requests a communications path be established to the called device. It contains the address of that called device and often the address of the calling device, too, so that the recipient knows where the call has originated.

Packet switch 4 looks at its routeing table to decide over which link the call request should be passed en route to its destination. It decides

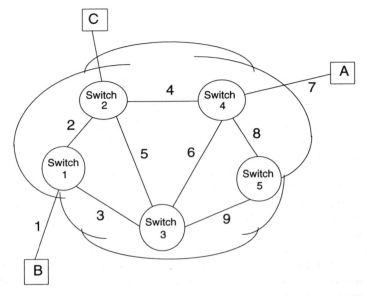

Figure 3.11 Addressing example

to pass the request to packet switch 2 which carries out the same process and sends it to switch 1 which passes the request to the destination device.

If B decides to accept the call request, it returns a call accepted packet along the route set up by A's request. The communications path is now successfully established, the virtual call created, between devices A and B and end-to-end information can now be transferred.

Should B reject the call, it sends a clear request packet to tell A that the call is unacceptable. At the end of the virtual call, a clear request packet is also sent by one of the parties and acknowledged with a clear confirmation packet.

3.4.3.1 *Logical channel numbering*

The network layer can support multiplexing, sustaining many active virtual calls simultaneously. For example, it can set up virtual calls to devices B and C from device A at the same time. As a result, a mechanism has to exist to identify the relevant virtual call along which packets are transmitted and received.

A unique Logical Channel Number is therefore assigned to each virtual call and used to refer to that virtual call for the lifetime of the call.

Each packet contains the LCN of the logical call to which it refers at its start, followed by a packet type field explaining the packet's function and any additional information that may be needed depending on the packet type.

For device A to set up a call to B, it assigns a unique LCN to the virtual call, carried in the call request packet.

The LCN indicates to packet switch 4 that all other packets transferred between switch 4 and A for this virtual call must carry the same LCN.

The call request packet is then routed across the network to packet switch 1, the most appropriate to device B. Switch 1 then assigns a unique LCN for the for the circuit to device B and sends the call request packet to B using this new LCN, which means all other packets for this virtual call must carry this LCN.

Usually, as in the above example, the LCNs at each end of the virtual call will be different. Furthermore, the LCN is only significant between an end point device and the packet switch to which it is connected.

It is important to remember that within the PSN, internal network links may be subject to different, often enhanced, protocols from those affecting the link between a switch and its attached end-point devices.

Just as in a conventional telephone system, a normal subscriber is plugged into the network via standard circuitry, whereas sophisticated communications equipment comes into play, from fibre optics to satellites, beyond that subscriber interface. At the other end, of course, will probably be another conventional telephone.

The role of X.25 as a standard interface to and from PSNs is paramount in presenting a common face to a user, to whom the complexities and richness of the internal PSN are transparent, and to an end-point device from the attached packet switch.

3.4.3.2 *Network Layer flow control*

Clearly, the sequence of information transfer at the Network or packet level is similar to that of the Data Link Layer. Error detection and correction are by and large left to layer 2, but the Network Layer does impose its own flow control mechanisms. Layer 2 flow control affects all traffic on the link; Network Layer flow control operates only on the information relating to a single virtual call running over a link which may be supporting numerous other virtual calls.

If the Data Link Layer is flow controlled and data is temporarily blocked from running over the circuit, then no Network Layer packets can be transferred. If the Data Link Layer is not flow controlled, but an individual virtual call is being blocked, other virtual calls can operate normally, unaffected by the temporary blocking of that single virtual call.

Flow control at layer 3 is governed, as at layer 2, by the window N system, typified by the X.25 level 3 protocol. The standard sequence number for data transfer in each direction takes up three bits in the packet, with possible sequence numbers running from zero to seven. The maximum window size is therefore seven packets before acknowledgement is needed.

Changing the window size in the Network Layer has a similar effect to changing it in the Data Link Layer, only more so. In a virtual call from device A to B, every packet from A has to pass through packet switches 4, 2 and 1 before reaching B. In other words, the packet on this route must pass across four different links.

Each link adds delay (transit time) before the data gets through. For each packet a network layer acknowledgement must follow down the same tortuous path in the opposite direction to reach device A.

If, for instance, each link in the virtual circuit above imposes a delay of 10ms, ignoring for this example any delay within the switches themselves, the four links used between end point devices A and B amount to an imposed delay of 40ms on each packet travelling one way, and double including the obligatory acknowledgement. But the delay is proportional to packet length and the acknowledgement is usually short, unless piggy-backed.

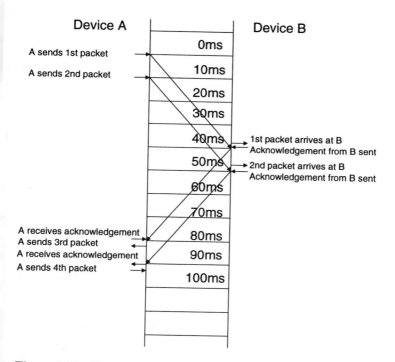

Figure 3.12 Illustration of a window size of two

A window size of one means device A only transmits one packet every 80ms: a maximum packet rate of 12.5 packets every second.

A window size of two makes a dramatic difference. Device A sends the first packet then another 10ms later, the minimum delay time (Figure 3.12). It takes 40ms for the first packet to reach B, for which B sends an acknowledgement. The second packet arrives 10ms later and is also acknowledged. Acknowledgement of the first packet arrives at device A after 80ms, time taken for the round trip. A then sends a third packet, receives acknowledgement of second packet 10ms after the arrival of the first acknowledgement and is then able to send a fourth packet into the network.

In other words, device A sends two packets in every 80ms period: 25 packets a second to B, twice the rate of a window size of one. If a window size of seven was imposed, device A could send seven packets before the first could feasibly be acknowledged and achieve a packet rate of 87.5 each second.

It may seem odd, then, that the most common window size used in PSNs with the most common interface, X.25, is only two, so that the maximum number of packets that can remain unacknowledged at any one time in each direction is two. Cynics have pointed out that X.25 equipment is usually delivered with the window size set to two and is never altered, but a significant factor is also the storage buffer needed to support a larger window.

At any time, any device may be called upon to buffer the entire window size: all packets sent inside the window of a virtual call. Moreover, that packet switch or end point device may be supporting numerous virtual calls simultaneously, putting considerable potential pressure on its buffer space. If the buffer fills, the device has to resort to emergency action which may mean taking down some calls to retain others.

The Network Layer in some PSNs allows for different window sizes depending on the nature of the virtual call, mainly those calls that need maximum throughput. It will depend on the delay time imposed by different links and packet switches en route between the end point devices. The optimum window size will vary according to the end device being called. In other circumstances, an average or compromise window size is chosen to accommodate most calls.

3.5 Datagram networks

An alternative to the concept of the Switched Virtual Circuit or virtual call in providing communications between end point devices over a PSN is a datagram network.

The phases of call set up, information transfer and call clearance are not needed in a datagram network. Each packet is simply launched into the PSN, leaving it up to the PSN to route the packet to its destination. Without call set-up there is no need for LCNs in the

Network Layer, but each packet must therefore contain its source and end point destination addresses.

Communicating devices in a datagram network are not connected to each other as in a virtual circuit, so the packet switches do not have to keep records of active calls and can therefore be kept both simple and fast.

If the amount of data to be sent can be confined to a single datagram packet transferred between devices, it can also be very efficient. As call set-up of a virtual call is one of the most time consuming operations for such networks, if the call period is only short, but set-up rates are high, the PSN performance can be seriously impaired.

Described as a lean and mean method of shifting data, datagram bearer networks place responsibility for error correction, resequencing packets on delivery and lost packets from the PSN to the end point node or user device. In terms of the OSI model, these functions are left to the Transport Layer, one above X.25.

Data is broadcast into the network and left to fend for itself, sometimes described as 'spray and pray', bringing with it some disadvantages:

1. There is no flow control mechanism in a datagram network comparable to that of a virtual call network.
2. As each datagram packet contains its explicit addresses and the network contains no record of connections, every packet switch en route has to look up the specific address in its routeing table to find out where it should switch that packet. This can prove slower than having a LCN identify a route, as happens in a virtual call network.
3. There is no formal acknowledgement system. In circumstances in which errors may be among the data stream due to adverse line quality and so on, a sender cannot be sure the datagram packet was successfully received.

After much controversy and changes of mind, the ITU-T X.25 1984 recommendation removed any support for the connectionless atagram method. For many network managers it is comforting to be able to track data to its destination.

Datagrams, however, can still be effective internal network protocols. The idea is to combine the speed and simplicity of datagram packet switches with the advantages of a virtual call network, producing a topology in which the domain outside the cloud is connection oriented, but inside is connectionless. Perhaps the largest example is the US Department of Defence's military oriented research network, ARPANET.

3.6 Routeing

A packet entering a PSN cloud from an end point device has to be routed to its destination device by the packet switches within the network according to various techniques, each with its own advantages and drawbacks. The choice is usually a balance between the complexity of the solution and the breadth of routeing functionality provided.

The most common PSN architecture is an irregular mesh, reflecting the gradual, if not actually haphazard growth of any large PSN aiming to meet changing user demands. A routeing strategy must be flexible enough to cope with a different topology as the need arises.

Switch manufacturers provide many different routeing solutions; researchers have produced literally dozens of routeing algorithms or rules such as 'hot potato', 'selective flooding' and 'delay- based, single-path distributed adaptive per-packet' routeing. They are all decision making processes for deciding which output link is most appropriate for a given data packet, each allied to a type of routeing.

3.6.1 Fixed or static routeing

The most common overall routeing choice is between fixed, sometimes called static or directory routeing, and dynamic routeing, with some variations between the two.

The simplest and least expensive strategy is fixed routeing, in which the network's packet switches are provided with routeing tables containing all the information they might need to route packets

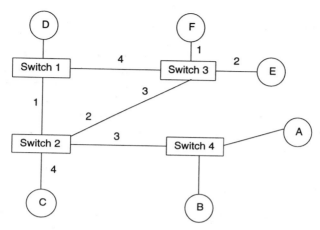

Figure 3.13 Routeing example

Table 3.1 Routeing table for Switch
2 of the example given in Figure X.13
                    3

End point device 2 address	Link
A	3
B	3
C	4
D	1
E	2
F	2

over that network. It is fairly common among X.25 PSNs to have
fixed routeing.

A routeing table for a packet switch comprises entries for each end
point device connected to the network. In Figure 3.13 there are six,
so the routeing table for switch 2 in this example (Table 3.1), has six

Table 3.2   Routeing table for Switch 3 of
the example given in Figure 1.13

End point device 3 address	Link
A	3
B	3
C	3
D	4
E	2
F	1

entries, each showing the address of the end point device (see also Figure 3.11) and the immediate link down which packets intended for each device should be routed.

A routeing table for switch 3 (Table 3.2) reveals a different picture because it looks at the end point devices from a different perspective within the network.

Packets for more than one end point device can be routed down the same link. For switch 2 this is true for end point devices A, B and E, F, in which case, there will be at least one more switch en route to those destination devices.

A prerequisite of fixed routeing is the construction of routeing tables which are then loaded into the network switches, usually when the network is first configured. Adding a new end point device to the network means updating the table in each switch, a potentially burdensome task for managers of medium to large networks.

Fixed routeing is straightforward, but depends on manual building and upkeep of the tables, for which a detailed knowledge of the network topology and traffic loadings is necessary. Software table building programmes are available to ease this task.

Not all PSN links necessarily have the same capacity and an understanding of the particular network usage can be essential for efficient communications. For example, in Figure 3.13 there are two routes between switch 2 and switch 3, one direct, the other indirect through switch 1. The direct circuit may, in fact, be slower than the alternative or the traffic may be heavy enough to warrant splitting it between the two routes. Traffic for device E through switch 2 might best travel directly, while traffic for F should take the indirect route.

A move in the direction of dynamic routeing is domain addressing, in which end point device addresses contain some routeing information. In a network of four switches, each switch can be regarded as a separate domain, from one to four. Addresses of end point devices linked to each might have a digit referring to their immediate domain, followed by digits identifying them individually within that domain, such as 101 (number one device linked to packet switch 1,) or 304 (number three device linked to switch 4).

Routeing tables in domain addressing are kept to a minimum; not every end point device need be recorded as many may share the same link to a domain. The switch needs to know only where the domains are to be found, rather than every device within those domains. Instead of updating every routeing table on attaching a new device, only the table of the switch to which the device is directly linked needs to be altered.

Also, table search time is reduced when a call request packet arrives and a virtual call path needs to be set up. Alternatively, in a datagram network, as each datagram packet has to be routed individually, any routeing time saved is particularly significant.

Large PSNs even create sub-domains within domains, just as the digits in an orthodox international telephone number can be sub-divided into national, regional and individual areas. As a packet switch only needs routeing information about its immediate domain, the domain addressing technique can be practical for large PSNs without the burden of enormous routeing tables.

However, static routeing is poorly equipped to cope with line failures. As the routeing tables are fixed, the link will remain unavailable until it is restored. Even though there might physically appear to be more than one route from A to B, packets and virtual calls cannot suddenly re-routed if the normal route fails.

As a result, network management systems capable of informing managers of any untoward events as soon as possible are of paramount importance, as manual intervention to alter the routeing tables is the only option.

In practice, load sharing between multiple routes is often a partial answer, with automatic fall backs in the case of a network failure. Alternatively, switches can refer to secondary and even tertiary routeing tables if the primary options become unavailable, although it demands more skill on the part of the network administrators.

Fixed routeing with some degree of dynamic alteration in the event of node failure is generally common in commercial networking environments.

3.6.2 Dynamic routeing

Dynamic routeing is a complex solution favoured in academic or networks such as ARPANET in which security or the lack of commercial demands favour more esoteric methods.

Rather than refer to predetermined routeing tables, packet switches capable of dynamic routeing make their own routeing decisions based on prevailing network conditions. Network managers and installers therefore do not need such a detailed knowledge of the live network; the implementation looks after itself to a degree.

The algorithms involved are capable of recognising both predictable network changes, such as time of day variations in network traffic and unforeseen alterations, such as intermittent circuit problems or a power outage. They are able to compute the optimum path through the network under prevailing conditions, even if it applies for only a short period, such as to relieve congestion at a particular switch.

A switch can time stamp delays on a link, for example, by sending packets known as ping packets to neighbouring nodes and measuring the time needed for the ping packet to complete a round trip.

Where circuit availability between user devices is paramount, dynamic routeing offers immediate alternative routeing between end points in the event of a line or switch failure. Also, load sharing of traffic between routes is more easily obtained, maximising circuit usage and minimising packet transit delay, if decisions can be made by the switches on a packet-by-packet basis.

If some links in the network belong to one or more public PSNs, load sharing through dynamic routeing can ensure that circuits paid for according to traffic load are optimised for maximum cost effectiveness.

3.6.3 Distributed adaptive routeing

A version of dynamic routeing assigns a cost value to each circuit based on available bandwidth, link loading, transit delay and throughput. The number of links or hops in the route is also assigned a value. The optimum path provides the overall lowest cost between end devices on that specific occasion; congestion at a node, for instance, may well increase the cost so that a cheaper route is found for the rest of the virtual call.

Packet switches can exchange route status information at regular intervals with adjacent or more distant nodes to build up a picture of current network conditions a capability known as distributed adaptive routeing. ARPANET, for instance, employs this form of routeing and is able to circulate information regarding link availability and route loading estimates to end point devices.

Routeing tables in these type of switches contain an entry for each end point device attached to the network. When the network is first installed, these entries contain only estimates of route loadings to the end devices.

The network's aim is always to make use of the optimum route between communicating end point devices: a route in which the transit delay is as low as possible. Once in action, the network estimates the delay of any one route by combining its own information regarding the size of the packet queue with data from the switches to which it is directly connected.

At defined intervals a switch sends a delay table to every switch to which it is directly connected. This contains an entry for every end point device on the network. On receipt of a delay table, a switch combines it with its own current routeing table, producing a new table.

For each end point device, each switch compares the delay in its routeing table with that indicated in the received delay table, adjusted for the effects of the link on which it was received. If the new delay is shorter than that for the route currently in use, the routeing table entry for that end point device is updated to show a better route has been found.

The result of this operation is that advice of alterations in network loading can be transmitted automatically through the PSN. As the size of each routeing table is itself dynamic, any new end point devices added to the network automatically appear in the switch tables soon after initial connection. The location of every end point device linked to the network is therefore available at any time within the network.

Any new circuit added to the network will automatically be taken into account in computations where using it results in shorter packet transit delays.

The overall ability of a PSN with dynamic routeing to adapt itself dynamically to changing fortunes and topology gives it remarkable operational independence and strength. Monitoring devices whose location and status are known at all times is straightforward. As no manual changes are involved when expanding the network, extra links or end devices can be added easily.

Delay table transmission between switches can be fixed according to line speed. Most of the time, nothing has changed, so that large networks can generate a considerable transmission overhead, wasting switch and circuit capacity. A solution is to send delay tables only when events deem it necessary: when something in the network has changed.

A new route, a new active end point device, increased delay time on a route, could all constitute an event in such a strategy. A stable, constant network will therefore not be burdened with unnecessary delay table transmissions.

Datagram PSNs are particularly suitable for DAR as no information about virtual calls needs to be kept within the switches and the volume of network related information on the move is therefore manageable. In contrast, the data necessary to decide how and when to switch packets is found within the switches on the route of a virtual call; if packets are to take a different route, the virtual call information also has to be moved.

DAR for a virtual circuit network is best provided by running the virtual call protocol above the datagrams. The datagrams convey the virtual call information between end point devices, while only the end point devices need to know anything about the virtual call protocol. BT's international data network operates in this way.

3.7 Packet switching equipment

An example of a successful packet switching system is DPN-100 from Northern Telecom. As well as private corporate networks over 30 PTTs have installed DPN networks as the basis of their national X.25 services. DPN switches are modular, so that they can be purchased to match the capacity requirements of the network, and subsequently grow with future needs. In addition to X.25, a variety of access protocols, (all major international and most de facto standards such as SNA) are supported, enabling the support of a multi-vendor data processing environment. Trunks can be directly attached to the DPN-100 at speeds of up to 1.544Mbit/s (US T1) or 2.048Mbit/s (UK MegaStream and equivalent). Enhancements to the DPN range include the support of Frame Relay as an access protocol.

A major advantage of the DPN-100 range is its conformance to international and industry networking standards including X.25 as an interface to packet mode devices, X.31 for access to ISDN packet-mode services, X.32 for PSTN and circuit-switched public data networks accessed by X.25 devices, X.75 gateways and X.3,X.28 and X.29 for asynchronous devices.

The range also provides transparent communications between otherwise incompatible protocols and devices, enabling users to choose equipment from multiple vendors, unrestricted by proprietary vendor protocols. The protocols at either end of a resulting DPN-100 network do not have to be the same to communicate. Examples include SNA/SDLC, 3270 display system protocol for IBM 3270 bisynchronous terminals, and asynchronous polled interface (API) for relevant point-of-sale terminals.

The switch and network architecture focuses on users' need for transparent access to applications, shielding them from control and security mechanisms necessary to maintain the integrity of the data passing across the network.

The DPN-100 range gives the network controller the ability to define network service targets based on user needs: fast application access, no interruptions during application processing and fast system response. There is also an emphasis on secure information flow through the establishment of Closed User Groups and Network User Identifiers, plus passwords as desired. It also highlights the importance of comprehensive network management, providing a real time network control system with dynamic updates of switch and overall network performance at configurable central, regional and local levels.

4. Fast packet switching

4.1 Introduction

The telecommunications industry is currently being driven by switching technology which is changing at an ever increasing pace. The demand for high capacity systems is becoming increasingly important for providing both sophisticated networking applications and offering flexible new services.

Today's carrier systems, E1 (2.048Mb/s) in Europe and T1 (1.5Mb/s) in the United States, provide a popular choice for networking voice traffic and data employing Time Division Multiplexing (TDM) techniques. This is due mainly to the functionality and cost savings that intelligent TDMs offer, and the fact that there have not been any other alternatives; that is until Fast Packet Switching came onto the scene.

In today's corporate networks voice traffic often represents around 80% of a network's capacity while data is only about 20%. However, many data applications are growing at 30% to 40% per year, while voice applications are growing at only 5%. This change in focus means that a communications network must meet today's heavy voice demands, yet be able to adapt to increasing data demands. For example, with the introduction of Local Area Network (LAN) applications and ever more powerful client/server workstations, the large amounts of data which must be transferred will become a critical issue in a company's overall communications network.

Fast Packet Switching is sometimes regarded as similar to ITU-T X.25 packet switching, but the two techniques must not be confused, despite having a common name. Both methods take transmissions from user devices, partition them into packets and route them over a backbone of network switches until they arrive at the right destination. However, due to its protocol transparent nature a Fast Packet Switch (FPS) approach uses technology which can handle all types of transmission, not just data; as well as providing much higher transmission speeds within the 150 to 600 megabits per second range. Now, applications that depend on high bandwidth can co-exist on the same network, as well as the low speed (64kb/s and less) data applications addressed by traditional packet switched networks.

Fast Packet Switching is a new digital communications approach that is revolutionising communications networking strategies, and proprietary FPS implementations are already widely used in the US and gaining momentum internationally. There is however some confusion surrounding the term 'Fast Packet Switching' (Blum, 1995). It is usually taken as a concept covering Frame Relay and Cell Relay technologies, as illustrated in Figure 4.1.

Frame Relay uses a variable length frame, and as such creates variable delays across the network. It is a data-link layer approach

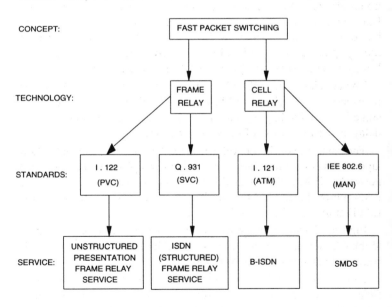

Figure 4.1   Fast Packet overview

(level 2 of OSI) and as such depends on the end systems supporting the Frame Relay protocol. These properties make Frame Relay best suited to non-delay-sensitive information transfer, such as data communications or still image.

Cell Relay on the other hand uses a fixed length frame or packet to carry information, and as such does not create variable delays across the network. It is a Physical Layer approach and consequently is protocol independent. These properties make Cell Relay suitable for transfer of delay sensitive information such as live speech.

4.2 Frame Relay

ITU-T recommendation I.122 — 'Framework for providing additional Packet Mode Bearer Services', describes the architectural framework for two types of Frame Relay service. The frames are based on Recommendation I.441. Frame Relay is proving to be the most popular of the additional packet mode bearer services defined. It also is the most simple (Mann, 1995; Kraemer, 1995).

The development of ISDN Frame Relay services is a packet mode interface to narrowband and broadband ISDN networks, and is designed to provide high speed packet transmission, minimal network delay, and efficient use of network bandwidth.

The basic aim of Frame Relay is to exploit the similarity of the OSI layer 2 function (data link control) between X.25 packet switched networks, IBM's SNA signalling protocol (LAPD), the IEEE 802 LAN protocols, and MAP/TOP. The standard (LAPB) frame and the LAPD frame are illustrated in Figure 4.2.

Figure 4.3 shows the Frame Relay communication paths relative to the OSI seven layer model. It is only at layer 3 that these protocols begin to look substantially different.

IBM originated what we now think of as the layer 2 or link layer function with its SNA SDLC protocol, and submitted it to various bodies for standardisation. SDLC emerged from ANSI as Advanced Data Communications Control Protocol (ADCCP), from ISO as HDLC and from ITU-T as LAP (the link access procedure for X.25). LAP evolved into LAPB for X.25 and into LAPD for ISDN. HDLC

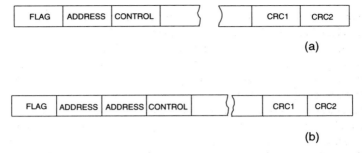

Figure 4.2 Frame Relay format: (a) standard LAPB frame; (b) LAPD frame as for Frame Relay

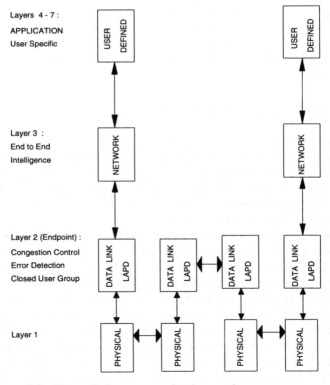

Figure 4.3 Frame Relay communications path

was adapted to the IEEE 802.2 Logical Link Control (LLC) for LANs, which also became the link layer for MAP/TOP.

In simpler terms, the object is to use a Frame Relay (level 2 routing) interface to connect between different types of communications networks because the link layers of these networks are already similar. The potential impact on corporate networking is significant, because the approach is relatively straightforward to implement compared to a layer 3 approach, and existing hardware can be utilised.

This idea is by no means a new one. It was suggested back in 1975 during the formative stages of the ITU-T X.25 recommendation for the packet switch network interface, that a simpler protocol would have many attractions. However the suggestion was rejected because of the proliferation of error-prone analogue transmission facilities still in existence at the time, requiring added complexity for error recovery. Digital transmission facilities are now widespread enough for a simplified protocol with error detection only, to be a viable option.

4.2.1 Frame Relay Networks

Frame Relay has been proposed to the ITU-T as an adaption layer protocol for connection to ATM networks. Although it is possible to implement Frame Relay interfaces on time division multiplexers (TDM) and packet switches to produce so called Frame Relay switches, Cell Relay platforms are by far the most efficient and flexible partner for the Frame Relay interface.

Some vendors implement a Frame Relay to X.25 interface on their packet switches, which only gives the performance of a traditional packet switched network. Time Division Multiplexers with Frame Relay interfaces have the disadvantages outlined below. A packet switch with a different software load removing the level 3 functions in the switch and creating fixed packet sizes on the switch trunks effectively achieves Cell Relay functionality for data traffic. There are also proprietary ATM multiplexers available which handle voice and data traffic, although standard ATM platforms are evolving (See Section 4.3 and Chapter 5). For the benefit of the following discussion, such platforms will be regarded as 'Frame Relay Switches'.

Instead of transferring bits between fixed locations like a circuit, a Frame Relay network acts like a wide area LAN. A sending device places an addressed frame into the network and it arrives quickly at its destination. To achieve these objectives, the network simply 're-lays' the packet, or frame, to a destination indicated by the layer 2 address field of the packet.

The Frame Relay switch performs the core layer 2 functions, of frame separation with flags, zero bit insertion, frame multiplexing via the address field, and CRC error detection to enable frames with errors to be discarded. The switch does not acknowledge or request re-transmission. This and all other protocol functions (layer 3 and up) are implemented end-to-end through the network, rather than by it.

Simplifying the protocol functions allows the network to operate cost effectively at high speeds and low delays. Instead of connecting switches at 64kb/s, it is done at N x 64kb/s (V.35 or X.21 interfaces) or with narrowband ISDN at 2Mb/s or 1.5Mb/s. Packets, instead of being processed at intermediate switches (as would be the case with conventional packet switches), are relayed directly from the originating switch to the destination switch. The result is that the delay incurred in relaying a packet is much less than the delay incurred in switching (software processing) a packet.

With the implementation of Frame Relay on a Cell Relay backbone, network bandwidth is used more efficiently. The pool of available bandwidth is shared among all applications (data, voice etc.). Bursty, high speed Frame Relay connections can access the entire network trunk bandwidth for very short intervals, and then release it for other applications. This unique property of Cell Relay switching makes high performance, bursty Frame Relay connections affordable because bandwidth does not have to be dedicated to those connections (as would be the case using a time division multiplexing backbone).

4.2.2 Benefits available with Frame Relay

LAN/WAN integration is by far the most popular application for Frame Relay services, although LAN to Host and still-image transfer are also possible.

Together Cell Relay switching and Frame Relay provide:

1. Savings on hardware. Only one port is needed with Frame Relay, since frames contain their destination addresses this effectively splits the physical connection into multiple logical connections to different destinations. This replaces the TDM interface requirement of dedicating one port for each destination. The result is reduced hardware costs, especially in large networks. Figures 4.4 and 4.5 illustrate this.

2. Bandwidth savings. Cell Relay switch handling of bursty data, when used with the Frame Relay interface provides continuous savings on line costs by offering more efficient management of bandwidth. Idle data is not transmitted over the network as it is with TDM platforms. Bandwidth is instantaneously used for bursts of data when necessary, and is then

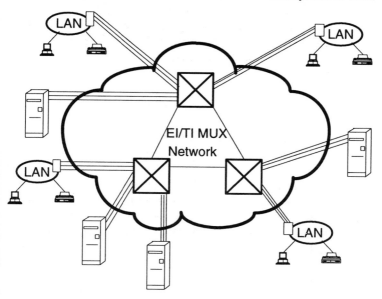

Figure 4.4 Conventional nework without Frame Relay

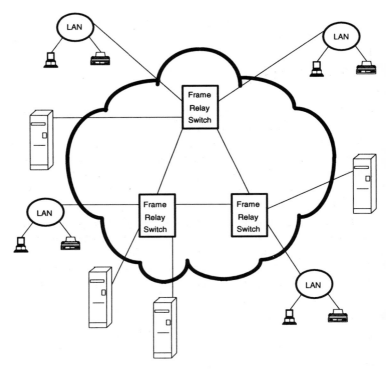

Figure 4.5   Network with Frame Relay

instantaneously freed for usage by other traffic. Bursts from multiple sources are interleaved on the same available bandwidth.

3.   Better performance. Cell Relay switching together with Frame Relay offers better performance than TDMs with Frame Relay. The Frame Relay interface can operate at much higher speeds without wasting bandwidth on the backbone. This results in lower delay and higher network performance. Because the Frame Relay interface provides full interconnectivity, every device is directly connected to every other device. This eliminates delay caused by routeing traffic through inter-

mediary (end system) nodes, such as LAN routers or front end processors.

4.3 Cell Relay

Cell Relay platforms transmit all information, including voice, data, video, and signalling in a single packet format. They provide truly integrated transmission over a single high-speed digital line. Unlike traditional packet switches, Cell Relay switches use short, fixed length packets (cells), and, using a hardware-based switching technique, switch them at very high speeds (100,000 to 1,000,000 packets per second), as in Figure 4.6.

Because Cell Relay networks have very high throughput and low delays, they can be used for all kinds of communication traffic: voice, synchronous data and video, as well as the low speed data applications that are being serviced by conventional packet networks to date. The use of a common packet format for transport of all network traffic results in simple packet routeing and multiplexing.

All packets are of the same length, use the same number of address bits, and are transported through the network using common switching, queuing, and transmission techniques, no matter what the connection type or its bandwidth requirements. Other than control information, the only real differences between packets are content and destination. Packets or cells are only generated on the trunk when actual information is present, therefore genuine bandwidth on demand is provided.

Comparing TDMs and packet switching with Cell Relay, the main benefits of traditional packet switching networks are lower costs, due to statistical multiplexing techniques, and higher reliability, due to the self- routeing nature of individually addressed packets. The main benefit of TDMs is that dedicating bandwidth provides the high throughput and low delays needed for voice, video and high speed data communication. Cell Relay switching delivers the best of both worlds by bringing the economy and reliability of packet switching to traditional TDM applications, such as voice, and the high throughput and low delays of TDMs to packet applications.

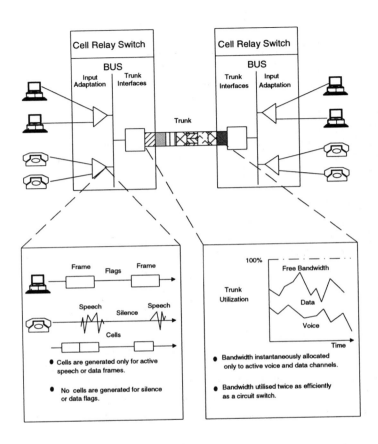

Figure 4.6 Cell Relay switch

Some manufacturers have introduced hybrid TDM and packet switched platforms as an attempt to capture the benefits of both technologies in one product, as in Figure 4.7.

Hybrid switching attempts to provide both circuit and packet switching features by dividing the slots on a TDM frame between voice channels and data channels. Since the voice slots are time

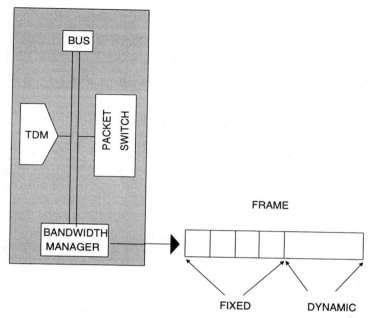

Figure 4.7 Hybrid switch

division multiplexed, and the data slots are packet switched, voice and data traffic must be separated to different sections of the switch.

With the simplest scheme, the voice/data boundary and frame length are fixed. This fixed boundary is inefficient because voice and data slots cannot be interchanged to accommodate statistical fluctuations in voice and data traffic. For more efficiency, the voice/data boundary can be designed to be movable, and the frame length dynamic, but not without complicated network analysis and increased complexity of switching architecture.

Hybrid switching is not really an integrated solution because not only are two different switches built, but a bandwidth management system is required that can deal with two separate types of data. In effect, there are two separate networks piggybacked onto one trunk. The product can cost twice as much as a truly integrated Cell Relay approach, with less of the simplicity and bandwidth economies to show for it.

There are two recognised branches of Cell Relay technology, Asynchronous Transfer Mode (ATM) and Distributed Queue Dual Bus (or Queued Packet Synchronous Exchange). Let us deal with these each in turn.

4.3.1 Asynchronous Transfer Mode

ITU-T recommendation I.121 Asynchronous Transfer Mode (ATM), is one of two approaches defined in Broadband ISDN. The other is Synchronous Transfer Mode (STM), which is an extension of traditional TDM principles providing fixed bandwidth channels and a packet switched signalling mechanism. ATM is a Cell Relay technology and offers the major advantage of flexibility over STM.

For ATM, standards bodies ANSI and ITU-T have agreed to use a 53 byte cell, consisting of a 5 byte header and 48 byte information field. The momentum to converge to ATM standards is evident in the adoption of a 53 byte cell format by the IEEE 802.6 committee for Metropolitan Area Networks.

ATM is a wide area networking approach which is why it is more suited to B-ISDN. The network platforms using ATM will be in the form of switching multiplexers or digital cross-connect switches. The intention is that ATM networks will act as the backbone for LAN, MAN and other existing networks, consequently there are adaption layer protocols being defined to provide the necessary re-formatting of frames and packets to the ATM Cell format.

Figure 4.8 shows the Broadband ISDN protocol model for ATM.

True standards based ATM networks are not anticipated to be available until well after Frame Relay and MAN approaches. Major vendors world-wide are developing ATM switches with a close eye on standards. Those with proprietary switches are developing modifications to meet the standards. The major force driving B-ISDN in Europe is the Deutshe Bundespost which is due to end a Broadband ISDN applications trial (Berkhom project) in 1992. The platforms used for the trial are STM and proprietary ATM switches. In the United States AT&T has had a 160Mb/s proprietary switch on trial for

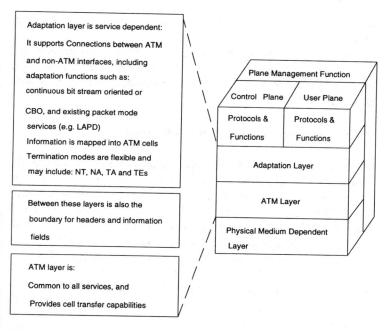

Adaptation layer is service dependent:
It supports Connections between ATM
and non-ATM interfaces, including
adaptation functions such as:
continuous bit stream oriented or
CBO, and existing packet mode
services (e.g. LAPD)
Information is mapped into ATM cells
Termination modes are flexible and
may include: NT, NA, TA and TEs

Between these layers is also the
boundary for headers and information
fields

ATM layer is:
Common to all services, and
Provides cell transfer capabilities

Plane Management Function

Control Plane | User Plane

Protocols & Functions | Protocols & Functions

Adaptation Layer

ATM Layer

Physical Medium Dependent Layer

Figure 4.8 B-ISDN protocol model for ATM

over a year. Public services are not expected to be widely available until well into 1996. However private networks are currently available using proprietary ATM approaches and standards compliant private ATM networks.

4.3.2 DQDB Metropolitan Area Networks

MANs are based on ring or bus topologies and are intended to provide city-wide networking, to interconnect LANs and to carry digital voice and video. MANs are also seen as a means of accessing broadband ISDN (ATM) networks in the longer term, hence the alignment between MAN and ATM Cell formats, (i.e. 53 byte cells).

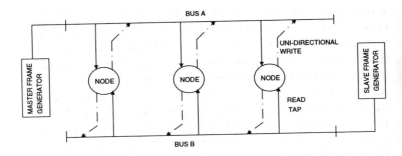

Figure 4.9   DQDB architecture

The standard IEEE 802.6 Distributed Queue Dual Bus MAN comprises two contra-directional data buses up to 150km long running up to 140Mb/s, as shown in Figure 4.9.

The bus may be looped into a ring format to give a much more resilient topology than LAN ring approaches. The ring is not closed, in the way a token ring LAN is. It is configured such that information terminates at the loop closure point on each side of the 'opening'. If a node or cable segment fails the logical opening is moved to the physical outage, and operation continues without degradation.

DQDB MAN technology is used in the Switched Multimegabit Data Service (SMDS) currently being used in the United States by some of the Regional Bell Operating Companies (RBOCs). SMDS is designed for access speeds of 1.544Mb/s and 45Mb/s. In Europe ETSI is developing standards for a similar service and some PTTs have already installed MANs.

Although SMDS is based on MAN technology, it will be possible to link the MANs together later using T3 lines to create Wide Area Networking. SMDS is a connectionless network unlike ATM which will be connection oriented.

4.4 Conclusions

It is generally accepted that the network of the 1990s will be some kind of extremely fast and completely integrated network, combining voice, video, image, data, signalling and high speed LAN transfers onto one backbone. The fast packet concept addresses these applications and uses the next generation communication technologies for future integrated broadband communications networks, which will be capable of handling all narrowband and broadband services.

Frame Relay services have already been announced by major carriers to provide an alternative to traditional leased lines. These virtual leased lines offer bandwidth in the way bursty applications require it, and charge for usage in a similar way to X.25 services. However all the benefits of Frame Relay apply. Private Frame Relay networks are also in use.

Although MANs, in the form of SMDS, are very visibly supported in the US by Bellcore and the Regional Bell Operating Companies (RBOCs) in terms of market development and technology life cycle, MANs are at a much earlier stage than Frame Relay. While the RBOCs are pushing SMDS for city-wide networking, standards have not been defined for interconnection of these networks to form wide area networking. The RBOCS are also actively engaged in Frame Relay trials.

International standards bodies have concluded that Asynchronous Transfer Mode provides the only effective backbone technology for bursty broadband communications over a wide area, such as LAN to LAN connectivity and digitised video/image distribution. Developments to date have underwritten this.

LAN routers with Frame Relay interfaces connecting into an ATM backbone (B-ISDN) along with voice services (VPN/ISDN) is one model for a private network. Another is LANs connecting into MANs, which are in turn linked by dedicated high speed WAN circuits or again into an ATM backbone with voice services on the MAN or on the ATM backbone.

4.5 References

ANSI (1990) *TI.606, Frame Relaying Bearer Service — Architectural Framework and service description.* ANSI Inc.

Blum, E.P. (1995) New developments in high speed networking technologies, *Spectrum Information Systems Industry*, Decision Resources Inc., 24 May.

CCITT (1988a) *Recommendation I.122, Framework for providing additional packet mode bearer services.* Blue book, ITU, Geneva.

CCITT (1988b) *Recommendation I.121, Broadband aspects of ISDN.* Blue book, ITU, Geneva.

Hullet, John L. and Evans, Peter (1988) New proposal extends the reach of MANs. *Datacommunications* (February)

Kanzow, Jurgen (1991) *The Berkhom Project.* IEE Review (March)

Kreamer, H. (1995) Evaluating Frame Relay platforms, *Stratcom White Pater*, June.

Krautkremer, T. (1994) Frame Relay: a natural evolution, *Telecommunications*, November.

Loe, D. (1994) Carried by word of mouth, *Communications Networks*, May.

Mann, K. (1995) Getting Europe in the frame, *Network Europe*, June.

McGibbon, A. (1994) Has Frame Relay dropped the bottom? *Communicate*, February.

Miller, A. (1994) From here to ATM, *IEEE Spectrum*, June.

Modahl, Mary A. and McClean, Karyn P. *Frame Relay's Impact.* Forrester Research Inc. Network Strategy Report.

Mollenauer, Jim (1989) The Global LAN is getting closer. *Datacommunications International* (December)

Moran, K.L. (1990) Fast Packet and Frame Relay. *Communications Magazine* (September)

Moses, J.T. (1993) Fast ethernet: an evolutionary alternative for high-speed networking, *Telecommunications*, August.

5. Asynchronous Transfer Mode

5.1 Overview and content

Current communications networks fall into discrete types dependent on the types of traffic being handled. Voice networks have been based on the use of circuit switching which offers guaranteed path availability and uniform delay during a connection. Network congestion results in the blocking of calls rather than delay. Data services often possess a very bursty traffic pattern and packet networks, with their flexible mechanism for interleaving traffic from different sources, offer an efficient solution. Variable delays, however, are introduced into the path as a result of queueing and these, though acceptable for data, would not be tolerable for voice and other constant bit-rate (CBR) services.

The increasing use of optical transmission and the general availability of large-scale integrated digital technology, offer the possibility of an integrated approach to handling all types of traffic source, including a range of new services, some, such as video-based services, requiring much broader channel capacities.

It is against the background of this Broadband Integrated Services Network (B-ISDN) that the technique known as ATM is being defined.

5.2 History

Circuit switched techniques such as those used for voice communications have been used in combination with STDM (synchronous time division multiplex) techniques. Here, the bandwidth on a link is divided into channels consisting of (typically) eight bits. The identity of a particular channel is defined by its relationship to a fixed frame reference. In contrast, in packet switching a block of data is associ-

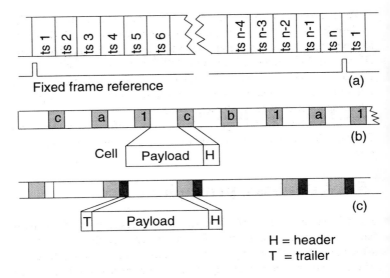

Figure 5.1 ATM compared with STDM and packet mode: (a) synchronous time division multiplexing; (b) Asynchronous Transfer Mode; (c) packet mode

ated with a 'header' which both defines its length and contains an address defining its identity.

ATD (asynchronous time division), arguably lies somewhere between STDM and packet switching and is intended for the transport of data, video and voice services. (Figure 5.1) Versions of the technique emerged in Europe in the 1980s in which a bit-stream was sub-divided into short, fixed-length slots (32 bytes maximum) with each slot having an address field attached to it. The term 'cell' was used to describe the slot plus the address field and transmission links would carry a contiguous stream of fixed length cells.

The work on ATD was a major influence when CCITT (now ITU-T) Study Group XVIII commenced the definition of the target transfer mechanism for B-ISDN in 1988. This transfer mechanism was to be standardised for use at the User/Network interface for B-ISDN and to be used as the basis for switching and multiplexing within B-ISDN nodes.

Asynchronous Transfer Mode (ATM), the name used by SG XVIII to describe the chosen solution, owes much to the early ATD proposals but some features were also adopted from Fast Packet Switching (FPS), a technique developed in the US aimed at reducing the end to end delays inherent in X.25 packet switching. FPS still used variable length packets but had much reduced protocol complexity to permit a high speed hardware implementation.

ATM employs fixed length cells having 48 information bytes and 5 header bytes which contain address information and other functions.

5.3 ATM basic principles

Asynchronous Transfer Mode is a specific packet oriented mechanism for the transfer of digital information based on the use of 'cells' which are of constant length, having a payload field and a 'header'. The cells are transmitted contiguously on a transmission link and are not identified by their position in relation to a fixed time reference but by means of address information in the header defining a 'virtual channel'.

The technique is *asynchronous* in the sense that the cells carrying a particular address (i.e. within a particular virtual channel) may appear at irregular intervals within the cell-stream. The technique is connection oriented in that a 'virtual circuit' is established at call set-up time and this will associate the virtual channels used on a series of network links to form the end to end connection.

ATM offers a flexible transfer capability common to a broad range of services with widely varying traffic patterns and can be employed on various transmission media operating at widely varying rates. ATM adaptation functions are provided to accommodate the data formats and operating characteristics of specific services.

5.3.1 The ATM cell

The cell consists of a header (label of 5 bytes) and a payload (information field of 48 bytes).

The header includes (Figure 5.2):

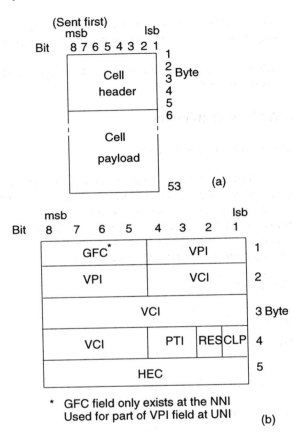

Figure 5.2 The ATM cell: (a) cell structure; (b) cell header structure

1. Address field defining the virtual channel to which the cell is assigned, divided into two parts, a virtual path identifier (VPI), and a virtual channel identifier (VCI).
2. Payload type identifier (PTI).
3. 8-bit CRC field for header error control (HEC). This field also supports the mechanism for cell delineation, i.e. the identification of the start of each cell within a serial bit stream.

The payload may be user information (voice, data, images, etc in digital form), signalling, or O&M messages (operations & maintenance).

The payload is transferred transparently from end point to end point across the network.

5.3.2 ATM multiplexing

5.3.2.1 *Multiplexing mechanism*

Information from a particular user is assembled into cells as it becomes available (Figure 5.3). Cells from a number of such sources will be placed in a queue in their order of appearance and the cells on the queue output will be placed in order in contiguous slots on the transmission link. If the aggregate information rate from the set of sources being multiplexed exceeds the link capacity at a particular

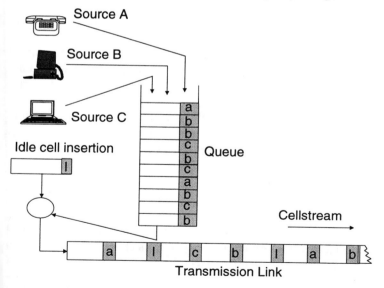

Figure 5.3 ATM multiplexing

time, then the queue fill will become greater, increasing the delay in the system or even causing loss if the queue becomes full. If source activity is low, such that the queue becomes empty, then 'idle cells' are inserted on the link to maintain the contiguous cell flow.

The sequence of cells from a particular user or source all carry the same value in the header address field and this cell sequence constitutes a 'virtual channel': the rate of transmission of cells within the virtual channel can be variable, reflecting the source activity and resource availability and it is in this sense that the transfer mode is said to be asynchronous.

5.3.2.2 *Physical layer*

An ATM multiplex is capable of being supported on a range of possible physical media and mapped into a number of existing bit-stream structures. For B-ISDN two possibilities are envisaged at the User-Network interface (UNI): a cell-based interface where timing is derived from the cell-stream and OAM messages are carried in special OAM cells with a unique header; a Synchronous Digital Hierarchy (SDH) based interface where the cell stream is mapped into the C-4 container and then packed into the VC-4 virtual container along with the VC-4 Path Overhead (POH). The ATM cell boundaries are aligned with the STM-1 octet boundaries and the H-4 pointer indicates the next occurrence of a cell boundary, offering a supplement to the HEC Cell delineation mechanism (Figure 5.4).

5.3.3 ATM switching

5.3.3.1 *Switch function*

The cell header address field is sub-divided into two parts: the virtual path identifier (VPI) and the virtual channel identifier (VCI, as in Figure 5.5). The virtual path identifier field offers the possibility of establishing semi-permanent virtual paths. These would be established in cross-connects controlled via network management and may be used for private circuits, for access network paths to particular service providers or for semi-permanent virtual trunks. Each of these

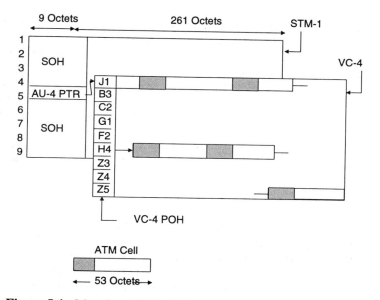

Figure 5.4 Mapping ATM cells into the SDH STM-1

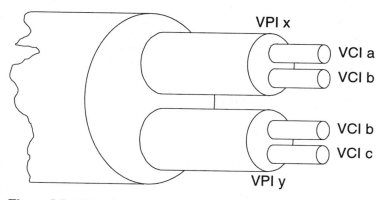

Figure 5.5 Virtual paths and virtual channels. VCIa and VCIb represent two of the possible values of VCi within the VP link with the value VPIx. Similarly VPIx adn VPIy are values within the physical layer connection. (Based on Fig 1/I.150)

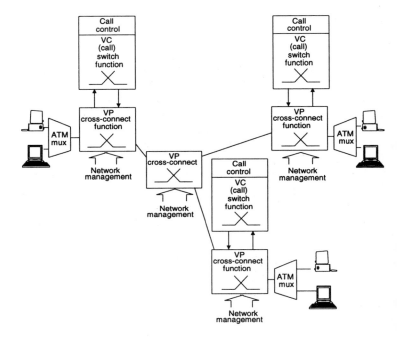

Figure 5.6 Example of a VP and VC switching hierarchy

could carry a number of virtual channels which, in the case of semi-permanent end-end paths could support a set of user defined processes. In the case of access paths and trunks, the virtual channels would provide the basis of switched connections handled at switching nodes. (Figure 5.6).

5.3.3.2 *Switch mechanism and characteristics*

A switching node, whether it be a VP cross connect or a VC switch, will have a number of incoming and outgoing ATM multiplex streams and the function of the switch is to transfer cells with a particular address field code on a particular incoming multiplex, to a particular outgoing multiplex (space switching) whilst placing the appropriate code in the cells' address field for use at the next switching node (asynchronous time switching). This function is performed according to a translation stored at the switch control and maintained for the duration of the 'virtual connection' (Figure 5.7). Before cell switching can take place, cell delineation is performed on the incoming multiplex cell streams, to permit extraction of cells and analysis of their headers. The 'idle cells', inserted into the multiplex at times of low source activity, are now removed and it is an important feature of ATM switching that the matrix is not burdened with handling these cells: the switch handles only the real traffic from the links.

The transit time of an ATM switch will depend in the main on the traffic levels on the queues providing access to the outgoing links and may vary considerably with time: the queues have to be dimensioned such that the probability of cell loss due to overflow is acceptably small.

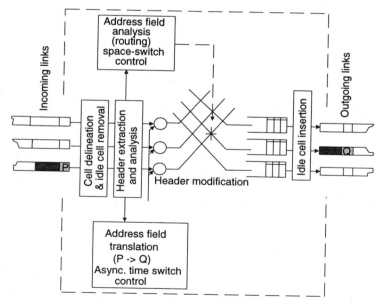

Figure 5.7   Functional representation of an ATM switch

5.3.3.3 *Switch implementation*

There are many possible implementation architectures for the internal structure of an ATM switch node, the choice being influenced by factors such as maximum capacity, growth and modularity requirements, connection types to be handled (point to multi-point connections present specific problems) and general performance issues. Specific architectural issues include:

1. The topology of the overall structure: ring, bus, single/multi-stage array etc.
2. The method of routeing used within the switch matrix: Self-routeing or translation table controlled.
3. The location of the buffer memories within the switching elements.

5.3.3.4 *Topology*

Dependent on the required characteristics, an ATM switch can be implemented using a variety of possible topologies. There are two basic parts of the structure, the port and the interconnect, and topology is determined by the means of providing the interconnect between the ports. Some examples of topologies are:

1. Bus, ring: the simplest switching arrangement where a number of ports are 'hung' on a single bus and communicate using a defined protocol which may use slots or some form of contention resolution. Simple and easily expanded but limited in overall capacity by the throughput of the bus. A ring is in effect a closed bus offering the possibility of different access protocols but with generally similar limitations.
2. n × m matrix: an array of simple 'on/off' switch elements providing the possibility of many co-existent paths between n inputs and m outputs (analogous to the electro-mechanical crossbar switch). Avoids the capacity limitations of the bus and ring by spreading the traffic over a number of paths but

expansion of the number of ports can result in very large cross-point matrices.

3.   Multi-stage matrices: If switching elements with a p × q capability are used instead of the simple 'on/off' function, then more complex arrays may be constructed having a variety of possible characteristics. In general, such networks may be expanded by addition of stages to handle very large numbers of ports, but do not expand in total size as rapidly as the simple matrix when additional ports are added; also they usually offer a multiplicity of path possibilities between each pair of inputs and outputs and this offers advantages in connection with both traffic and security. The individual elements of such a matrix may themselves be implemented using topologies such as those described above. Batcher-Banyan networks and Clos networks are examples of such multi-stage matrices.

5.3.3.5 *Routeing*

Self-routeing switches require only one translation at the matrix input and this results in the appending of a header prefix, used for routeing purposes by each of the matrix switching elements, and discarded on exit from the switch. In translation table controlled switches the path is marked locally in translation tables within each switching stage, permitting routeing to proceed on the basis of the original external header. Routeing through the switch matrix may be connection oriented i.e. cells always follow the same route through the switch and therefore remain in sequence. Alternatively, a connectionless approach to routeing within the switch may be taken, the output port address being carried in the cell and individual cells being routed toward that port according to the local routeing algorithm at the particular point in the switch matrix. With connectionless routeing, cell sequence integrity across the matrix may not be maintained.

5.3.3.6 *Buffering*

Cell buffering in the switching elements can be organised on three possible bases:

1.   Input buffering: One buffer dedicated to each switching element input. Contents can be switched to one (or several in point to multi-point operation) output.
2.   Output buffering: One buffer dedicated to each switching element output, with no blocking of cells from any of the switching element inputs.
3.   Central buffering: one common buffer memory block is provided for all cells in the switching element.

5.3.4 Traffic control and resource allocation

5.3.4.1 *Admission control*

It is anticipated that the ATM based B-ISDN will carry a mixture of traffic sources varying from uniform rate to very bursty, from very short to very long duration. When a new connection is requested it is necessary to ensure that capacity can be allocated on the network resources such that the new connection will provide the required Quality of Service (QoS) and that the QoS targets of existing services are still met. This is done by expressing each requested connection in terms of a set of Flow Enforcement Parameters (FEPs) and on each link through the network, an admission control algorithm decides whether or not to accept the connection. (FEPs could be based on estimates of the average and peak bit-rates and expressions of the burstiness.)

5.3.4.2 *Flow conditioning*

Having 'admitted' a connection, to ensure that actual cell flows conform to the parameters negotiated during connection admission, flow conditioning procedures are required. Without these, the QoS of other connections could be disrupted.

Flow conditioning has two components: flow enforcement (otherwise 'Police function' or 'Usage parameter control') which is mandatory and located just within the network periphery and flow throttling, which is optional and located between the source and the network boundary. (Figure 5.8, Andersen, 1990). Throttling is applied by the

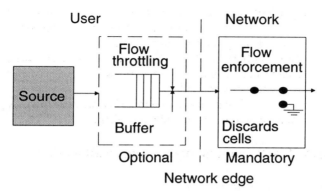

Figure 5.8   Flow conditioning functions

user to ensure that the source data flow conforms to the agreement. Enforcement discards cells which are in excess of the negotiated levels.

5.3.5 Key Advantages of ATM

The basic multiplexing and switching characteristics described above give rise to some of the most important advantages of ATM:

1.   Flexibility.
2.   Inherent adaptation between information flow rates and transport resources.
3.   Ability to handle a range of user traffic source rates using the same technique and equipment.
4.   Statistical multiplexing capability.
5.   Flexibility in handling the mix of service demand.
6.   Transport infrastructure is general purpose, being conceived to handle by adaptation, a very wide range of traffic/service types.
7.   Possibility of service integration on the same medium, an important consideration for a broadband network where service types and demand are still ill-defined.

5.4 Standards for ATM

Standardisation of broadband started in CCITT during the 1985–1988 study period which, by its close had resulted in the choice of ATM as the transfer mode to be used at the broadband User/Network Interface (B-UNI) and as the basic switching mechanism for all B-ISDN services, this being defined in the first version of recommendation I.121.

A set of standards for Broadband ISDN is being established in CCITT within Study group XVIII during the current study period (1989 – 1992) under the I-series recommendations and 13 recommendations dealing with ATM parameters are being processed under accelerated procedures.

Inputs to this process derive, amongst other sources, from the regional standardisation organisations (ETSI in Europe and T1 committee in North America) and from the major European Community communications R&D initiative known as RACE (Research in Advanced Communications technologies in Europe).

Other CCITT Study Groups became concerned with broadband, for example SG XI is active in considering the signalling requirements.

However SG XVIII remains the main focus and the following sections outline the contents of the 13 draft recommendations relevant to ATM, highlighting the key points and providing detail on the primary features (CCITT, 1990).

The content is, of course, provisional and subject to change as the process of standardisation continues: many of the paragraphs are annotated 'for further study'.

The structure of the standards is shown in Figure 5.9. Under a top-level document giving an overview of B-ISDN, there are five groups of documents covering:

1.   ATM functions.
2.   Broadband services.
3.   Network aspects.
4.   User-Network interface.
5.   Operations and maintenance.

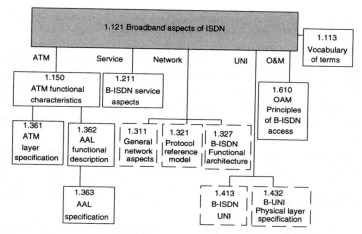

Figure 5.9 ITU-T I-Series standards for B-ISDN

5.4.1 I.113: Vocabulary of terms for broadband aspects of ISDN

Glossary of terms and definitions, organised under two headings: services; and interfaces, channels and transfer modes. These are kept updated to remain in line with the evolution of terminology used in the other recommendations in the series dealing with broadband.

5.4.2 Broadband aspects of ISDN

Statement of the basic principles of the broadband aspects of ISDN.
Establishes ATM as the transfer mode for implementing B-ISDN and identifies certain specific advantages:

1. Flexibility of network access.
2. Dynamic bandwidth allocation on demand: fine degree of granularity.
3. Flexible bearer capability allocation; easy provision of semi-permanent connections.

4. Independence from the means of transport at the physical layer.

5.4.3 I.150: B-ISDN ATM functional characteristics

Describes the basic principles of ATM and addresses the functions of the ATM layer:

1. Describes how the physical layer connection can carry a number of *virtual paths* distinguished by the 'VPI' field: also how each virtual path can carry a number of *virtual channels* distinguished by the 'VCI' field. (Figure 5.5.)

2. Defines a *virtual channel connection* (VCC) as a concatenation of *virtual channel* links that extends between two points where the adaptation layer is accessed. VCCs may be switched or semi-permanent, provide a Quality of Service defined by parameters such as *cell loss ratio*, *cell delay variation* and preserve cell sequence integrity. VCC traffic parameters will require negotiation and VCC usage will be monitored. (Figure 5.10.)

3. Describes how VCCs may be established and released citing four methods at the UNI: by subscription (semi-permanent connection set up by management); using a meta-signalling VC to establish a signalling VC; by using a signalling VC for user/network signalling; similarly for user/user signalling.

4. Defines a *virtual path connection* (VPC) as a concatenation of *virtual path links* that extends between the point where VCI values are assigned/translated/removed. VPC characteristics are broadly as for VCCs except that VPCs are only set up on a semi-permanent basis via network management.

5. Defines a number of uses for pre-assigned VPIs.

It deals with the definition of the following:

1. Cell multiplexing and switching: The VC is the basic routeing entity for switched services and is handled in VC multiplexers

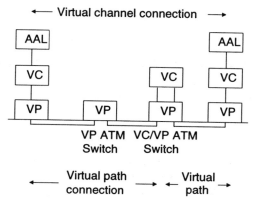

Figure 5.10 Types of ATM layer connection
(Based on Fig. 2/I.150)

and switches. VCs are aggregated in VPCs which are routed through VP multiplexers and switches.

2. Quality of Service related to VP and VC connections: particular QoS classes will be requested at call establishment for both VP and VC connections. A VPC will carry VCs of various different QoSs. The VPC must meet the most demanding QoS requirement of the VCs carried.

3. Quality of Service related to cell loss priority (CLP): many variable bit rate (VBR) services will require a guarantee of some minimum capacity as well as a peak capacity. When congested, the network will need to know which cells may be discarded without violating (e.g.) the guaranteed minimum capacity. The CLP bit is set by the user or service provider to indicate such cells.

4. Payload types (user information/network information) and the payload type identifier field.

5. Generic flow control (GFC) mechanism for control of user information towards the network at the broadband user-network interface (UNI) in order to alleviate any short term overload conditions that may occur.

5.4.4 I.211: B-ISDN service aspects

Provides a general classification of the standardised services to be supported by a B-ISDN and gives guidance on important network aspects which need to be taken into account when supporting services for B-ISDN.

In particular, gives consideration to video coding aspects in relation to ATM.

5.4.4.1 *Service classification*

The classification separates services into two main categories: interactive and distribution services. Interactive services are sub-divided into:

1. Conversational, which permit bilateral communication with real-time, end-to-end information transfer from user to user.
2. Messaging, which permit user-to-user communication via storage units with store and forward, mailbox and/or message handling functions.
3. Retrieval, which permit retrieval of information stored in information centres (usually) provided for public use.

Distribution services are sub-divided into those:

1. Without user individual presentation control, and includes broadcast services. Continuous information flow from central source to authorised receivers attached to the network. User has no control over start and order of presentation.
2. With user individual presentation control, information from central source to large number of users, provided as sequence of information entities with cyclic repetition. User can control start and order of presentation.

The recommendation deals with each class in detail, describing the type of information, giving examples of the services and their possible applications and defining some of their attributes.

5.4.4.2   B-ISDN aspects with an impact on services

The recommendation also gives guidance concerning some important aspects which need to be taken into account when supporting and developing services for B-ISDN, and also provides an introduction to recommendations I.362 and I.363 dealing with the ATM adaptation layer.

Multimedia aspects:

Deals with the need to separate out connection control from call control so that one call can support several types of connection dealing with different (standardised) information types.

Quality of Service aspects:

Deals with QoS negotiation and indication options, also the need for the Cell Loss Priority indicator.

Service bit-rate aspects:

1.   Constant bit-rate services: bit-rates negotiated at call set-up time to ensure resources are fully allocated for duration of call.
2.   Variable bit-rate services: expressed by a number of parameters related to traffic characteristics (see Rec. I.311). Parameters negotiated at call set-up time are supported for duration of call.
3.   Maximum service bit-rate supported by the 155 Mbit/s interface: UNI (SDH) payload = 155.52 x 260/270 = 149.760 Mbit/s. With ATM cell format, cell payloads carry 149.76 x 48/53 = 135.631 Mbit/s (= maximum possible service bit-rate). May in fact be less due to: transfer capacity required for OAM and signalling cells; ATM adaptation layer overheads; the time period associated with the 'structure' attribute for CBR services.
4.   Maximum service bit-rate supported by the 622 Mbit/s interface: for further study.

Service timing/synchronisation aspects:

1.   End-to-end methods: use of an adaptive clock; use of a synchronisation pattern; and time stamping.
2.   Network methods: mechanisms to be provided to support services with 8kHz integrity. Examples are network sourced time-stamped cells; timing information from the T-interface.

Connectionless data service aspects:

Can be supported in B-ISDN:

1.   Indirectly via a B-ISDN connection-oriented service
2.   Directly via a B-ISDN connectionless service (for further study).

Interworking aspects:

Services normally available from narrow-band interfaces will also be available from broadband interfaces.

Signalling aspects:

A number of service-based signalling capability requirements are listed.

5.4.4.3   *Video coding aspects*

The final section of the recommendation deals with video coding aspects and argues for coordination of video coding studies with B-ISDN studies in order to minimise the number of video terminals needed to access a range of interactive and distribution video and still image based services. Minimising the number of coding techniques and matching them to network characteristics is key to this aim. Constant bit-rate and variable bit-rate coding are discussed as are a layered approach to video coding for service integration and the impact of the ATM network on video coding. The recommendation concludes that layered coding combined with variable bit-rate coding has advantages for video service integration and for utilisation of ATM network capabilities and therefore recommends that studies should be concentrated on these methods.

5.4.5 I.311: B-ISDN general network aspects

Describes a number of separate network related aspects of B-ISDN: networking techniques, signalling principles and traffic control and resource management.

5.4.5.1 *Networking techniques*

This deals with the layering principles used in defining the B-ISDN and the hierarchical relationship between the Physical Layer, the ATM layer and the layers above.

It defines an ATM transport network as being structured in two layers: ATM layer and Physical Layer. The ATM layer has two sub-layers, the Virtual Path (VP) sub-layer and the Virtual Channel (VC) sub-layer.

The physical layer transport functions are sub-divided into three levels: transmission path, digital section and regenerator section levels.

I.311 takes further the VC and VP concepts introduced in I.150:

1. Virtual Channel (VC): a generic term to describe a unidirectional communication capability for ATM cells.

2. VC Link: VC between two consecutive ATM entities where the VC Identifier (VCI) value is translated. A specific value of VCI is assigned each time a VC is switched in the network.

3. VC routeing: translation of VCI values between the incoming and outgoing links on a VC switch.

4. VC Connection (VCC): concatenation of VC links. Extends between two VCC endpoints (more than two for point to multipoint). Provided for user-user, user-network, network-network information transfer. Cell sequence integrity is preserved by the ATM layer within a VCC.

5. VCC endpoint: where cell information field is exchanged between the ATM layer and the user of the ATM Layer Service.

6. Virtual Path (VP): generic term for a bundle of virtual channel links. All VC links in the bundle have the same endpoints.

7. VP Link: unidirectional capability for transport of ATM cells between two consecutive ATM entities where the VP Identifier (VPI) value is translated. A specific value of VPI is assigned each time a VP is switched in the network.

8. VP routeing: translation of VPI values between the incoming and outgoing links on a VP switch.

9. VP Connection (VPC): concatenation of VP links. Extends between two VPC endpoints (more than two for point to multipoint). Provided for user-user, user-network, network-network information transfer. Cell sequence integrity is preserved by the ATM layer within a VCC.

10. VPC endpoint: where VCIs are originated, translated or terminated. When VCs are switched the VPC supporting the incoming VC links must be terminated first and a new outgoing VPC created.

The Physical Layer is also explored:

1. Transmission path: extends between network elements that assemble/disassemble the payload of a transmission system. Cell delineation and Header Error Control are required at the end points.

2. Digital Section: extends between network elements which assemble/disassemble a continuous bit or byte stream.

3. Regenerator Section: a portion of a digital section.

Figures 1,2,3 and 4 of I.311 represent:

1. The relationship between VC, VP and Transmission Path (Figure 5.11).

2. The hierarchy of the ATM transport network (Figure 5.11).

3. The hierarchical layer to layer relationship (Figure 5.12).

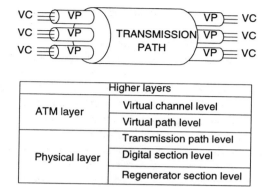

Higher layers		
ATM layer	Virtual channel level	
	Virtual path level	
Physical layer	Transmission path level	
	Digital section level	
	Regenerator section level	

Figure 5.11 Relationship between VC, VP transmission path (Fig. 1/I.311) and hierarchy of the ATM transport network (Fig. 2/I.311)

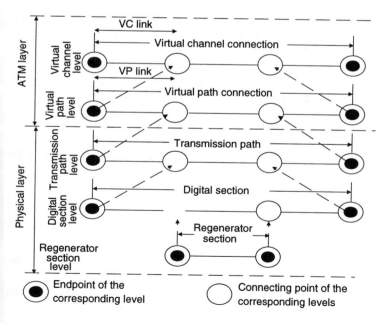

Figure 5.12 Hierarchical layer to layer relationship (Fig. 3/I.311)

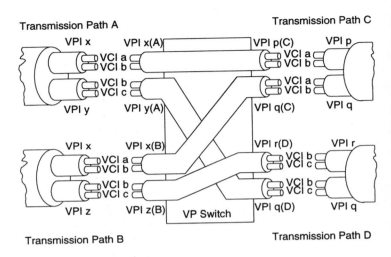

Figure 5.13 Representation of VP switching (an elaboration of Fig. 4/I.311)

4.   VC and VP switching (Figures 5.13 and 5.14).

I.311 explores the identified applications of VCCs and VPCs, categorising each as user-user, user-network and network-network. In each application the connection end-points are localised and an indication of possible use given.

5.4.5.2 *Signalling principles*

This identifies the signalling capabilities required in B-ISDN to support a multiplicity of service types and covers the requirements for establishing signalling communication paths. It examines the capabilities required to control ATM VC and VP connections for information transfer and (implicitly acknowledging the distinction between connection and call) further examines the capabilities required to support simple multi-party and multi-connection calls. Capabilities include:

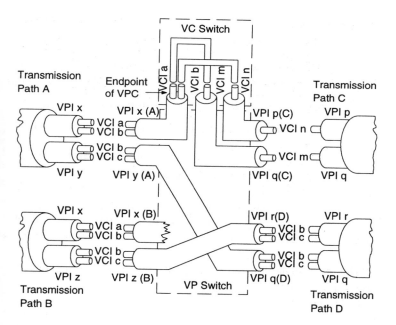

Figure 5.14 Representation of VP and VC switching (an elaboration of Fig. 4/I.311)

1. Establish, maintain and release VCCs and VPCs. May be on-demand, semi-permanent or permanent, but must comply with requested connection characteristics.
2. Support point-to-point, multipoint and broadcast configurations.
3. Negotiation of traffic characteristics at set-up and during calls.
4. Support of symmetric and asymmetric calls.
5. Addition/removal/reconfiguration of connections and parties from multi-connection and multi-party calls.
6. Support interworking between different coding schemes and with non B-ISDN services.

Signalling principles also examine the requirements for signalling virtual channels on the user access and in the network and also the requirement for meta-signalling procedures to establish, check and release signalling virtual channel connections.

Three types of signalling virtual channel are considered at the user access (those in the network are for further study):

1. Point-to-point, one in each direction allocated to each signalling end-point.
2. Selective broadcast, one allocated to each service profile (relates to supplementary services: see ITU-T Rec. Q.932). It can only apply in the network-user direction. A number of possible service profile configurations have been identified.
3. General broadcast, which is used for broadcast signalling independent of service profiles.

The meta-signalling is carried in a permanent virtual channel connection with a standardised VPI and VCI value.

I.311 illustrates some possible signalling configuration cases: each describes a possible provision of meta-signalling, signalling and information virtual channels and paths between the customer equipment (CEQ) and local or transit connection related functions (CRFs) or exchanges.

5.4.5.3 *Traffic control and resource management*

This deals with the several levels of traffic control capabilities to be provided by the ATM-based B-ISDN:

1. Connection admission control, which are actions taken by the network during call set-up phase in order to establish whether the call should be accepted or rejected. Acceptance is on the basis that sufficient resources are available to establish the call (at the required QoS) across the whole network without impact on the QoS of existing calls.
2. Usage parameter control, which are actions taken by the network during the information transfer phase of a call to ensure

that actual cell flows conform to the parameters negotiated during connection admission control, thus protecting the QoS of other calls on the network. (Also referred to as the flow enforcement or policing function.)

3.   Priority control, which is the possibility for the user to assign different priorities to different traffic flows by using the cell loss priority bit.

4.   Congestion control, which are mechanisms used by the network in the face of traffic overloads which prevent the network from guaranteeing the negotiated QOS to the calls in progress in the network.

5.4.6   I.321 B-ISDN Protocol Reference Model (PRM)

This is based on the ISDN PRM defined in I.320. This recommendation takes into account the functionalities of B-ISDN and defines a B-ISDN PRM, reflecting the principles of layering of communications as defined in the Reference model of Open Systems Interconnection (ITU-T Rec. X.200).

5.4.6.1   *Protocol reference model*

The B-ISDN PRM is represented as a cube with a number of horizontal layers and with three vertical planes: the user plane, the control plane and the management plane. (Figure 5.15.) The control plane and user plane are layered as follows: Physical Layer; ATM layer; ATM adaptation layer; higher layers. The Management Plane contains both layer related and plane related functions.

5.4.6.2   *PRM Layer Functions*

The basic functions in the PRM Layers are as follows:

1.   Physical Layer:
     Physical medium sublayer; bit timing; physical medium.

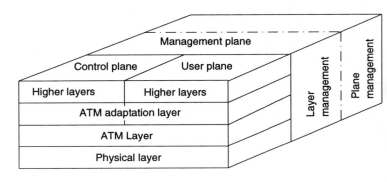

Figure 5.15   B-ISDN protocol reference model (Fig. 1/I.321)

        Transmission Convergence sublayer: cell rate decoupling; HEC header generation and verification; cell delineation; transmission frame adaptation; transmission frame generation and recovery.

2.   ATM Layer: generic flow control; cell header generation and extraction; VPI/VCI translation; cell multiplexing and de-multiplexing.

3.   ATM Adaption Layer:

        AAL Segmentation and Re-assembly sublayer (SAR); CS PDU Segmentation; CS PDU Re-assembly.

        AAL Convergence sublayer: convergence (service dependent).

5.4.7   I.327 B-ISDN functional architecture

This provides a statement of the basic functional architecture of B-ISDN to complement that for ISDN given in I.324. The architecture model defines reference configurations for B-ISDN, breaking down these into connection elements such as private access, public access, national transit etc. and identifying the locations of reference points for interfaces. The way in which functional groups relate to the connection elements is defined.

5.4.8 I.361 B-ISDN ATM layer specification

This defines the cell structure and the coding of the fields of the ATM cell header, and also addresses the ATM protocol procedures. (Figure 5.2.)

The cell header fields at the UNI are defined in Table 5.1, and the cell header fields at the NNI are defined in Table 5.2.

The coding of the fields is as follows:

1.   GFC field: 4 bits, UNI only. 0000 when not used. Coding for further study.

Table 5.1   Cell header fields at the UNI

Header field	Number of bits
Generic Flow Control (GFC)	4
Virtual Path Indicator (VPI)	8
Virtual Channel Indicator (VCI)	16
Payload Type (PT)	3
Cell Loss Priority (CLP)	1
Header Error Control (HEC)	8

Table 5.2   Cell header fields at the NNI

Header field	Number of bits
Virtual Path Indicator (VPI)	12
Virtual Channel Indicator (VCI)	16
Payload Type (PT)	3
Cell Loss Priority (CLP)	1
Header Error Control (HEC)	8

Table 5.3 Payload type identifier coding

Code	Payload type
000	User data cell; no congestion; SDU type 0
001	User data cell; no congestion; SDU type 1
010	User data cell; congestion; SDU type 0
011	User data cell; congestion; SDU type 1
100	VCC OAM F5 flow segment
101	VCC OAM F5 flow end to end
110	Reserved; future traffic control and resource management
111	Reserved; future functions

2. Routeing field: At the UNI = 8 bits VPI + 16 bits VCI. Specific values are assigned for meta-signalling virtual channel identification and general broadcast signalling virtual channel identification. At the NNI = 12 bits VPI + 16 bits VCI. Specific values are assigned for idle cell identification, physical layer OAM cell identification, code for the use of the physical layer, unassigned cell identification.

3. Payload type field: 3 bits for payload type identification. User information = 00. Payload type identifier field coding has been identified, as in Table 5.3.

4. Cell Loss Priority (bit): Value = 1: Cell is subject to discard depending on network conditions. Value = 0: cell has higher priority.

5. HEC field: operation described in I.432.

6. Reserved field: use not yet specified.

5.4.9 I.362 B-ISDN ATM Adaptation Layer (AAL) functional description

This classifies the services which may require AAL capabilities accessed through different Service Access Points (SAPs).

Table 5.4   ATM adaptation layer service classification

	Class A	*Class B*	*Class X*	*Class C*	*Class D*
End to end timing	Required		User defined	Not required	
Bit rate	Constant	Variable	User defined	Variable	
Connection mode	Connection oriented				Connec-tionless

The organisation of the AAL into two sublayers is defined. These are the Segmentation and Re-assembly (SAR) sublayer and the Convergence (CS) sublayer.

The Service classification is based on three characteristics:

1. Timing relationship between source and destination, needed or not needed.
2. Bit rate, constant or variable.
3. Connection mode, connectionless or connection oriented.

ATM Adaptation Layer (AAL) is required to support five classes of service, as in Table 5.4, where:

A	Constant bit rate (DS Service emulation)
B	Variable bit rate (VBR voice and video)
X	Connection oriented user defined services
C	Connection oriented services for data
D	Connectionless services for data

5.4.10 I.363 B-ISDN ATM Adaptation Layer (AAL) specification

This describes the interactions between the AAL and the next higher layer and between the AAL and the ATM layer, and also AAL peer to

Table 5.5 ATM AAL types

	AAL1	*AAL2*	*AAL3*	*AAL4*	*AAL5*
Timing relationship between source and destination	Required		Not required		
Bit rate	Constant	Variable			
Connection mode	Connection oriented			Connectionless	Connection oriented

peer operations. It describes in detail the five types of AAL relating to the five classes of service defined in I.362, in each case addressing the functions which may be performed in the AAL to enhance the service provided by the ATM layer. In particular the SAR functions are addressed and the format of the SAR Protocol Data Unit (PDU) is defined for each type.

Five ATM types are defined, as in Table 5.5, where AAL5 is a new addition.

5.4.10.1 *AAL Type 1*

Services provided to higher layer:

1. Transfer/delivery of constant bit-rate 'service data units'.
2. Transfer of timing information.
3. Indication when information is errored or lost.
4. Transfer of structure information.

Functions of AAL type 1 are:

1. Segmentation and re-assembly of user information.

2.   Handling of cell delay variation.
3.   Handling of lost/misinserted cells.
4.   Source clock frequency recovery at receiver.
5.   Monitoring for/ handling of AAL Protocol control information bit errors.
6.   Monitoring for/handling of user information field bit-errors.
7.   Handling of cell payload assembly delay.
8.   Source data structure recovery at receiver.

Segmentation and re-assembly (SAR) Protocol Data Unit (PDU): Type 1 SAR-PDUs contain a header having a 4-bit sequence number field to detect lost or misinserted cells, also a 4-bit sequence number protection field.

One bit in the sequence number field can be used to indicate the existence of a convergence sublayer.

5.4.10.2 *AAL Type 2*

Services provided to higher layer are:

1.   Transfer of variable source bit-rate SDUs.
2.   As for AAL type 1.
3.   As for AAL type 1.

Functions of AAL type 2: all as for AAL type 1, items 1 to 7.

Segmentation and re-assembly (SAR) Protocol Data Unit (PDU): Type 2 SAR-PDUs contain a header having a sequence number field to detect lost or misinserted cells and an information type field to indicate Beginning of Message (BOM), Continuation of Message (COM), End of Message (EOM); also a trailer containing a Length Indicator (LI) field and a CRC code field.

5.4.10.3 *AAL Type 3/4*

The convergence sublayer (CS) has been subdivided into the common part (CPCS) and the service specific (SSCS). Different SSCS protocols to support specific AAL user services may be defined or the SSCS may be null.

1.   Message mode service: transports single ASL-SDUs or, using blocking/deblocking internal functions, multiple AAL-SDUs in one or (optionally using SAR) more SSCS-PDUs.
2.   Streaming mode service: transports one or more fixed size AAL-SDUs in one or (optionally using SAR) more SSCS-PDUs.

Both service modes may offer the following procedures:

1.   Assured operations: flow control and re-transmission of corrupted/missing CS-PDUs ensures delivery of correct AAL-SDUs.
2.   Non-assured operations: Integral AAL-SDUs may be lost/corrupted and will not be corrected. Options are delivery of corrupted AAL-SDUs and flow control on point-point ATM layer connections.

SAR sub-layer accepts variable length CS-PDUs from CS and generates SAR-PDUs with up to 44 octets of CS-PDU data.
SAR functions in AAL type 3/4 are:

1.   Preservation of SAR-SDU: achieved by identifying segment type and SAR-PDU payload length.
2.   Error detection: bit errors or lost/inserted SAR- PDUs.
3.   Multiplexing/demultiplexing of multiple CS-PDUs concurrently from multiple SAR connections over a single ATM layer connection.
4.   Maintenance of SAR sequence integrity.
5.   Partially transmitted SAR-SDU abort facility.

Segmentation and re-assembly (SAR) Protocol Data Unit (PDU): Type 3/4 SAR-PDUs contain a header having an information type (segment type) field, e.g. BOM, COM, EOM, SSM (single segment message), a sequence number field and a Multiplexing identifier field

(MID), also a trailer containing a length identifier field and a CRC code field.

5.4.10.4 *AAL Type 5*

Under study.

5.4.11 I.413 B-ISDN User-Network interface

5.4.11.1 *Reference configuration*

This gives the reference configuration for the B-ISDN User-Network Interface (B-UNI), defining the functional groups and reference points which apply (Figure 5.16). These are essentially the same as for Narrow-Band ISDN as described in I.411. but to clearly identify the existence of broadband capabilities, the letter 'B' is used in the labelling of all reference points and functional groups.

Functional groups in the B-ISDN reference configuration are: B-NT1, B-NT2, B-TE1, TE2, B-TE2, B-TA. Reference points are: T_B, S_B, R.

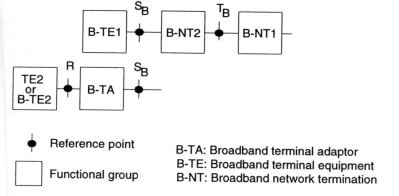

Figure 5.16   B-ISDN reference configurations (Figure 1/I.413)

5.4.11.2 *Physical realisation*

A limited set of examples of possible physical configurations is given. These cover configurations which could be supported by standardised interfaces at reference points S_B and T_B and also illustrate physical configurations for shared medium applications.

5.4.11.3 *Interfaces at the T_B and S_B reference points*

The characteristics of the interfaces at the T_B and S_B reference points are described. There are two physical layer options at T_B and S_B at 155.52Mbit/s and at T_B at 620Mbit/s: cell based and SDH based.

5.4.11.4 *B-ISDN model applied to functional groups*

A list of functions which might be included in each of the B-UNI functional groups is given, although these lists are not exhaustive:

1. B-NT1 functions: line transmission termination; transmission interface handling; OAM functions.
2. B-NT2 functions: adaptation functions for different media and topologies; functions of a distributed NT-2; cell delineation; concentration; buffering; multiplexing/demultiplexing; resource allocation; usage parameter control; adaptation layer functions for signalling (for internal traffic); interface handling (T_B and S_B); OAM functions; signalling protocol handling; switching of internal connections.
3. B-TE functions: user-user and user-machine dialogue and protocol; interface termination and other layer 1 functions; protocol handling for signalling; connection handling to other equipments; OAM functions.

5.4.11.5 *Physical layer information flows and interface functions*

Physical layer information flows between the physical medium (PM), the transmission convergence (TC) sublayer and their adjacent entities (ATM layer and management plane) are defined as are UNI related OAM functions.

5.4.12 I.432 B-ISDN User-Network interface, physical layer specification

This defines a limited set of Physical Layer interface structures to be applied to the S_B and T_B reference points. It also covers the physical medium and the transmission system structure that may be used at these interfaces and also addresses the implementation of UNI related OAM functions.

The main orientation of the recommendation is toward optical transmission since this is the preferred physical medium, however both optical and electrical interfaces are recommended. Implementations are to allow for terminal interchangeability. Maximum functional commonality between the UNI physical layer and that of the NNI is aimed at.

5.4.12.1 *Physical medium characteristics of the UNI*

Physical characteristics of the UNI at both 155.520Mbit/s and 622.080Mbit/s are considered, listing transmission media, interface range, electrical and optical parameters and connectors.

5.4.12.2 *Transmission convergence sublayer functions*

These are covered including the available information transfer rate, transmission frame adaptation functions (subdivided into 'Cell-based interface' and 'SDH based interface', and covering timing, interface structure and OAM implementation), header error control, idle (empty) cells and cell delineation and scrambling. From the ATM viewpoint, the mappings into SDH and the HEC field and cell delineation and scrambling algorithms are perhaps the most important.

1.   Transmission frame adaptation functions: SDH based interface. The interface bit-stream is based on SDH as described in ITU-T Rec. G.709. The ATM cell stream is mapped into the

C-4 container and then packed into the VC-4 virtual container along with the VC-4 path overhead. The ATM cell boundaries are aligned with the STM-1 octet boundaries. The C-4 is not an integer multiple of the cell length so cells will cross C-4 boundaries.

2. HEC field and cell delineation. The HEC field of the cell header is calculated across the entire cell header and the code used is capable of single bit error correction and multiple bit error detection. The 8-bit HEC field is the remainder of the modulo 2 division by the generator polynomial $x^8 + x^2 + x + 1$ of the product x^8 multiplied by the content of the header excluding the HEC field. Cell delineation is performed by using the correlation between the header bits to be protected and the relevant control bits introduced into the header by the HEC. In the 'hunt' state the delineation process is performed by checking bit-by bit whether the HEC coding law is respected (syndrome = zero) for the assumed header field. Once such an agreement is found, it is assumed to identify a cell boundary and the 'pre-sync' state is entered. In this state the process is confirmed several times before the 'sync' state is declared. In the 'sync' state, a certain number of incorrect recognitions of the coding law indicates loss of cell delineation causing transition to the 'hunt' state once more. (Figure 4.17.)

3. Scrambling will be used to improve the security and robustness of the HEC cell-delineation mechanism. In addition it helps randomise the cell data for possible improvement of the transmission performance. The self-synchronising scrambler polynomial $x^{43} + 1$ has been chosen for the SDH-based physical layer. A Distributed Sample Scrambler (DSS) using the polynomial $x^{31} + x^{28} + 1$ appears the probable choice for the cell-based Physical Layer. (ETSI, 1990)

5.4.13 I.610 OAM principles of the B-ISDN access

This covers the maintenance of the B-UNI and the B-ISDN subscriber access. It follows the maintenance principles defined in Rec.

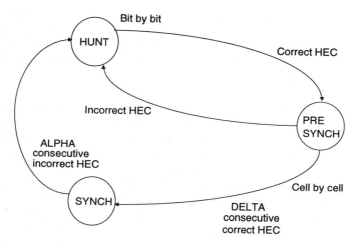

Figure 5.17 Cell delineation state diagram (Fig. 5/I.432)

M.20 and describes the minimum functions needed to maintain the Physical Layer and ATM layer of the customer access. Five phases have been assumed in specifying OAM functions for B-ISDN: performance monitoring, defect and failure detection, system protection, failure or performance information, fault localisation and the implementation of a Telecommunications Management Network (TMN) to support the operation of these phases is assumed.

OAM functions in the network are performed at five distinct levels of hierarchy associated with the ATM and Physical Layers of the PRM. Corresponding bidirectional OAM data flows result, F1 corresponding to the regenerator section level, F2 to the digital section level, F3 to the transmission path level, F4 to the virtual path level and F5 corresponding to the virtual channel level. The physical layer contains F1, F2 and F3 whilst F4 and F5 lie in the ATM layer. (Figure 5.18.)

The mechanisms to handle OAM flows are described. For physical layer flows, these are considered for three separate types of transmission system: SDH based, Cell based and PDH based: use of SDH path and section overheads, maintenance cells for the physical layer

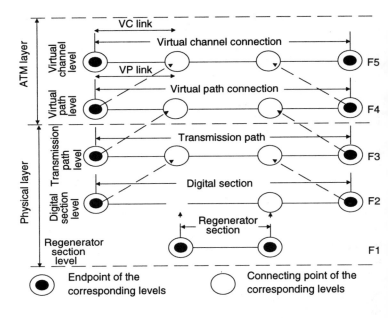

Figure 5.18 OAM hierarchical levels (FIg. 3/I.610)

and bit messages in PDH are described. For ATM layer flows, special cells dedicated to OAM functions are provided.

A tabulated approach is used to specify the OAM functions of the UNI and their related OAM flows. Tables are provided for the SDH-based physical layer, the cell-based physical layer and the ATM layer.

5.5 Quality of service performance requirements

Quality of Service (QoS) is a set of measures of service performance which determine the degree of satisfaction of a user of the specific service.

In an IBC network different services will have different QoS requirements. Voice traffic can tolerate only limited delay but can accept a moderate loss of cells. High speed data may be very sensitive to information loss but be relatively insensitive to delay. In services using coding techniques in which cell loss may result in loss of synchronisation, the time interval between cell loss events may be more critical than the actual cell loss. Thus QoS requirements have to be stated in the context of the specific bearer service concerned. Further, the user-oriented QoS requirements have to be related to specific network related network performance (NP) requirements before the requirements of a particular QoS on the network can be parameterised.

Network performance parameter values are required to give network providers a view of the performance they must provide to meet the users' needs and to provide manufacturing industry with performance targets upon which to base equipment designs. They are defined against the following assumptions:

1. All ATM connections maintain cell sequence integrity.
2. Each teleservice is supported by a single ATM bearer service.
3. Each ATM bearer service is supported by one virtual circuit.
4. ATM bearer services are regarded as different if at least one of their attributes differs.

The QoS of an ATM Layer service can be characterised by a number of parameters: cell insertion rate due to header errors; cell loss rate due to header errors; cell loss rate due to buffer overflow; mean/maximum delay; delay dispersion/burst diffusion; connection set-up delay; reliability/availability; information field error-rate.

These correspond to the attribute network performance for connection types and in the following section definitions are given for a set of network performance parameters based on the ITU-T attribute description method which characterises the requirements for a number of important and characteristic services. Some of the issues governing choice of parameter values are discussed and the values proposed for each bearer service are given in tabular form. (Tables 5.6 and 5.7.) This material is based on a literature survey and analysis

Table 5.6 Bearer services and applications

Bearer service	Application
64kbit/s 8kHz structured (CCITT Rec. I.231 196I.231.4)	Telephony; data; videotelephony
2x64kbit/s unrestricted, 8kHz structured (CCITT Rec. 1.231.5)	Videotelephony
< 64kbit/s	Access to server for connectionless services. These services include 'Remote process control'
64kbit/s	Interoffice signalling between signalling points in public networks or between two PABXs
384kbit/s unrestricted, 8kHz structured (CCITT Rec. 1.231.8)	Videotelephony/conferencing; data
1.92Mbit/s unrestricted, 8kHz structured	Videoconferencing; data
2.048Mbit/s	Circuit emulation (multiplex signal)
> 2Mbit/s up to < 130Mbit/s	Data; image; video
Approx. 1Mbit/s unidirectional	Stereo sound contribution
45–130Mbit/s unidirectional	TV contribution
8.448Mbit/s	Circuit emulation
34.368Mbit/s	Circuit emulation

made at the National Technical University of Athens as part of the RACE project 1014 'Atmospheric' (Anagnostou, 1991).

The network oriented QoS sub-attributes are sub-divided into two categories: information transfer related and call control related.

Table 5.7 Provisional QoS parameter values for bearer services

Bearer service	BER	CLR	CIR	End-to-end delay	Cell delay variation
64kbit/s	10^{-6}	10^{-4}	10^{-3}	400ms	20ms
2 x 64kbit/s				200ms	
Connection-less access interoffice signalling	10^{-6}	$3x10^{-4}$	10^{-3}	50ms	10ms
384kbit/s – 1.92Mbit/s	10^{-7}	10^{-5}	10^{-3}	200ms	20ms
> 2Mbit/s	10^{-7} with FEC(10^{-8} without FEC)	10^{-5} with FEC (10^{-9} without FEC)	10^{-3}	200ms	20ms
Stereo sound Broadcast TV	10^{-7} with FEC (10^{-8} without FEC)	10^{-5} with FEC (10^{-9} without FEC)	10^{-3}	500ms	20ms 10ms
8.448Mbit/s 34.368Mbit/s	10^{-7} with FEC(10^{-8} without FEC)	10^{-5} with FEC(10^{-9} without FEC)	10^{-3}	80ms	10ms

5.5.1 Information transfer parameters

These are as in the following sections.

5.5.1.1 Bit Error Ratio

This is the ratio of the number of bits incorrectly received to the total number of bits sent. In ATM it may be applied to the Cell information field only. It is expected to be no higher than the BER in present STM networks and will probably improve further with the introduction of optical transmission systems.

5.5.1.2 *Cell Loss Ratio (CLR)*

This is the ratio of lost cells in a given VC to the total number of cells entering the VC. Cell loss may occur due to errors in the header of the cell (misrouted cells) or due to buffer overflow. CLR is a key performance parameter in an ATM network and is an ATM specific attribute. Cell loss can be detected using sequence numbering of cells and this is regarded as the minimum adaptation layer functionality for all services.

Cell header errors affecting the destination address field could lead to lost cells with potential violation of service privacy and security, therefore such errored cells are discarded. Discard, however, can also have a deleterious effect on some services, particularly video where some coding algorithms may give error extension to a complete video frame. In this context, CLRs of 10^{-11} might be required and this is a very demanding target in terms of buffer lengths in the ATM network. Signals of higher digital hierarchies, 8 Mbit/s and 34 Mbit/s, may lose synchronisation if a cell loss affects justification control bits. Thus for all the high bit-rate services an FEC in the terminals for bit errors and cell losses may be necessary.

5.5.1.3 *Cell Insertion Ratio (CIR)*

This is the ratio of inserted cells to total number of cells entering a VC. It results from header bit errors in the address field. Cell insertion is considered more serious than cell loss: for some services it may cause loss of terminal synchronisation. FEC, whilst dramatically improving CLR, can actually adversely affect CLR by converting an errored header to a valid one but with the wrong VCI value. This effect can be tolerated only because CIR values are much less than CLR values.

5.5.1.4 *Cell delay variation*

This is the difference between the values of the transit delay of cells belonging to the same VC during a pre-defined period of time (tentative definition).

It is caused mainly by variations in cell delay in the network nodes, in particular in the queueing and cell rate adaptation buffers and can be augmented when asynchronous operation is used within the network node.

Delay variation can be controlled at the receiving terminal at the expense of increased delay and provision of greater buffer space. With a level of compromise between the requirements on terminals and switching centres, a maximum peak to peak delay variation of 100 microseconds per ATM switching centre can be expected, resulting in a network end-end delay variation of a few milliseconds.

5.5.1.5 *End to end transfer delay*

This is the one way propagation time between two S/T interfaces. It excludes Cell assembly/disassembly time and is not applied to lost/misrouted cells.

It is therefore confined to the delay caused by the transmission links and by the transit delays of the switches.

5.5.2 Call Control Parameters

These are as follows:

1.   Connection set-up delay, which is the interval between the event of the call set-up message transfer and receipt of its acknowledgment. (Called user response time is not included.) Provisional target maxima based on the I.352 figures for 64 kbit/s ISDN are:

Mean connection set-up delay of 4500ms.
95% connection set-up delay of 8350ms.

2.   Connection release delay, which is the interval between the event of the call release message transfer and receipt of its acknowledgment. Provisional target maxima based on the I.352 figures for 64kbit/s ISDN are:

Mean connection release delay of 300ms
95% connection release delay of 850ms.

5.6 References

Anagnostou, M.E., et. al. (1991) Quality of service requirements in ATM based B-ISDNs, *Computer Communications*, May.

Andersen, I., Sallberg, K., Stavenow, B. (1990) A resource allocation framework in B-ISDN. In *Proceedings of the X111 International Switching Symposium (ISS '90)*, **1**, p.111.

Batcher, K.E. (1968) Sorting networks and their applications. In *Proceedings of AFIPS 1968 SJCC, 1968*, pp. 307-314.

Bernstein, J. (1995) Clearing the air and muddying the ATM waters, *PC Week*, 21 February.

CCITT (1990a) Study group XV111 report R.34, (Geneva Meeting May), *Recommendations drafted by Working party XV111/8 to be approved in 1990.*

CCITT (1990b) Study group XV111 report R.45, (Matsuyama Session, November), *Recommendations to be approved in 1991.*

Clos, C. (1953) A study of non-blocking switching networks, *Bell System Technical Journal*, **32** (2) pp. 406-424.

Curtis, R. (1994) Introduction to Sonet ATM, *TI Technical Journal*, February.

Daddis, G.E. et. al. (1989) A taxonomy for broadband integrated switching architectures, *IEEE Communications*, May.

Dupraz, J., DePrycker, M. (1990) Principles and benefits of the asynchronous transfer mode, *Electrical Communication*, **64** (2/3).

ETSI (1990) Sub technical committee NA5 (broadband networks). *Draft CCITT Contribution: Distributed Scrambler*, Report on rapporteurs meeting, Lannion 22-26 October, pp. 79-85.

Fisher, D.G., Tat, N., Berglund, A. (1990) A flexible network architecture for the introduction of ATM. In *Proceedings of the X111 International Switching Symposium (ISS '90)*, **1**, p.105.

Gard, I., Rooth, J. (1990) An ATM switch implementation — technique and technology. In *Proceedings of the X111 International Switching Symposium (ISS '90)*, **4**, p. 23.

Goeller, L. (1994) What I always wanted to know about ATM, *Business Communications Review*, September.

Goldberg, L. (1995) ATM's growing pains bring maturity as UNI 4.0 evolves, *Electronic Design*, 30 May.

Haduong, T., Stavenow, B., Dejean, J. (1990) Stratified reference model — an open architecture approach for B-ISDN. In *Proceedings of the X111 International Switching Symposium (ISS '90)*, **1**, p. 105.

Hughes, D. and Hooshamand, K. (1995) ABR stretches ATM network resources, *Data Communications*, April.

Hui, J. (1988) Resource allocation for broadband networks, *IEEE Journal on Selected Areas in Communications*, **6** (9) Dec.

Keene, J. (1995) The current status of the ATM Forum's LAN emulation standard development, *Lightwave*, January.

Long, M. (1995) Meeting the challenges presented by ATM, *PC User*, 28 June – 11 July.

Ramakrishnan, K.K. and Newman, P. (1995) ATM flow control: inside the great debate, *Data Communications*, June.

Reeves, J. (1995) Low-speed access: extending the reach of ATM, *Telecommunications*, February.

Rickard, N. (1995) ABR: realising the promise of ATM, *Telecommunications*, April.

Schaffer, B. (1990) ATM switching in the developing telecommunication networks. In *Proceedings of the X111 International Switching Symposium (ISS '90)* **1**, p.105.

Seeley, A. (1995) Fast track, *PCLAN*, March.

Shelef, N. (1995) SVC signalling: calling all nodes, *Data Communications*, June.

Vincent, G. (1995) The superhighway in action, *IEE Review*, May.

Wernik, M.R., Munter, E. (1990) Broadband public network and switch architecture. In *Proceedings of the X111 International Switching Symposium (ISS '90)*, **1**, p.15.

6. Local and wide area networks

6.1 Introduction

Incredible though it may seem, a four-fold increase in data traffic is predicted by the year 2000, the beginning of the 21st century. Yet it is conceivable that this may be an underestimate when one considers the range of new multimedia services potentially available in the next few years, the ever growing power of computing resources and the continual imagination of business and residential applications in combining computer and communications as a single platform for services.

This chapter examines Local Area Networking (LAN) and Wide Area Networking (WAN), developments from the early 1960s, the phenomenal leaps of technology recorded during their development, the issues concerning implementing such networks today and the future role of such networks for business and residential communities.

6.2 Network development

In 1960 the term wide and local area networks was as foreign as MacDonalds. The wide area network was principally the transmission network implemented by the dominant monopolitic PTT/Telco, which was analogue based, with copper reaching to the subscriber and the backbone network pulled together by Frequency Division Multiplexing (FDM). (See Figure 6.1.)

The network was designed to carry analogue voice telephony, in the 300Hz to 3.4kHz frequency range. This arose out of the conversion of voice into an electrical signal and is still the vital component for all voice telephony. Frequency Division Multiplexing provided the means of grouping a number of 300kHz to 3.4kHz signals into a

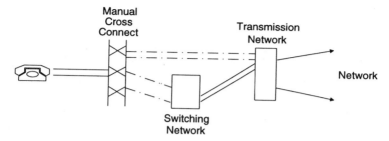

Figure 6.1 Analogue wide area network of the 1960s

single group by multiplexing them on to a higher and higher frequency carrier.

It allowed PTTs to build multiple groups of analogue circuits throughout their network with immediate consolidation of the raw transport mechanism, which at the time on the trunk network was coaxial cable.

In the late 1960s users began to build computer systems to assist in the management of their business. Initially these systems were centralised into one facility with clusters of user terminals and peripherals located around a large mainframe system such as IBM 3000 range. The benefits of computing applications were realised by such businesses as airlines, banking and insurance companies, so the pressure to remotely connect some of the user community from the mainframe became paramount.

The availability of modem technology allowed the mainframe applications to be distributed into remote sites over the analogue PTT network. (See Figure 6.2.)

Modem as the name suggests modulates and demodulates from analogue tone into digital data and vice versa. In practice it allowed digital data from a remote terminal to be converted into analogue signals for its onward connection into the restrictive (300kHz to 3.4kHz) analogue public network.

Earlier modems such as V22 (1200 baud) and V22 bis (2400 baud) offered, in comparison to modern modem, slow speeds. Nevertheless, these served to provide the initial revolution of Wide Area Network-

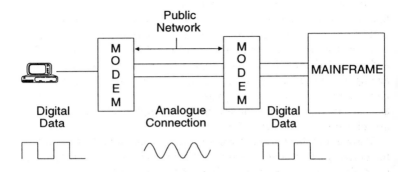

Figure 6.2 Remote mainframe applications over an analogue network

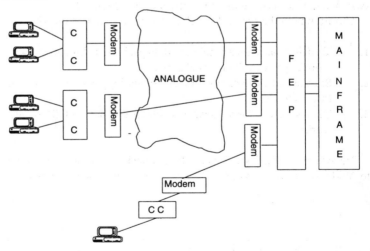

Figure 6.3 Distributed computing over an analogue network. (CC = Cluster controller; FEP = Front end processor.)

ing. Major analogue networks soon appeared as the platform for companies to distribute their centralised computing resources. Based principally around IBM's SNA environment they become the work-horse of many corporate organisations. (See Figure 6.3)

The heart of the service were analogue leased lines, as supplied by the monopolistic PTTs. As these networks increased and expanded better technology was developed to improve the speed at which data could be transmitted, with more sophisticated modems such as V.29 and now V.32 and V.42, and to optimise the manner in which the analogue service was presented using such techniques as multidrop and multiport. These had two goals, speed of information to the user, and to attempt to save money from the high PTT tariffing cost associated with such networks.

Throughout this period of growth other means were developed for the delivery of computer sources to the user which at first seemed a good alternative to the large analogue based networks.

The most notable was the introduction of X.25 networks. These networks combined leased services to the user at a local network with switched data services withn the network. (See Figure 6.4.)

Unfortunately the earlier PTT X.25 services had some drawbacks:

1. Tariffing based on a number of data packets, so high usage was more expensive than leased services.
2. Slow speeds, where new modem technology was now offering faster speeds.
3. Delays across large networks which did not suit SNA networks.

Nevertheless, X.25 services could be considered as an alternative to Wide Area Networking and would have important benefits for

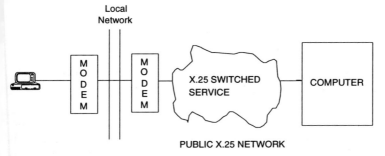

Figure 6.4   Remote computer access over the X.25 network

many network business applications. These included a wide connectivity capability, unlike leased services the switched capability enabled a one to many connection rather than a one to one. The ability to handle protocol conversion in its simplest form, such as different speeds, and a more robust networking service, since it was based on the PTT core service.

Modem networks grew phenomenally during the 1970s both domestically and internationally. The SNA networks offered efficient data connection based on synchronous data connection, however, many computer systems also adopted different protocols for communication, including asynchronous type connections. Synchronous connections essentially rely on the accurate timing of the two elements trying to communicate. This allows fast exchange of information, given that both elements are synchronised.

Asynchronous connection on the other hand assumed poor timing between the elements. Therefore data was surrounded by error checking and start/stop information.

Synchronous connection is much more efficient than asynchronous since large amounts of data can be transmitted faster. To improve the efficiency of asynchronous connections statistical multiplexing methods were developed during the late 1970s, both for local and wide area connectivity.

The principle of statistical multiplexing was to take advantage of slow data connections associated with asynchronous connections. This allowed a number of asynchronous devices to utilise one circuit for onward connection. (See Figure 6.5.)

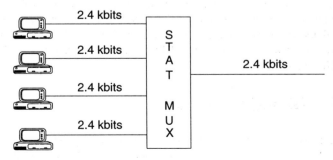

Figure 6.5   The user of statistical multiplexing

The statistical multiplexer's role is to handle the inward, and in reverse the outward, connections using buffering and retransmission techniques. This simple process provided incredible improvement in the efficiency of the time and combined with modem devices became the work horse for many asynchronous based analogue networks.

Voice services remained within the PSTN, although some international applications and large corporate networks were developed around analogue leased services. This is very true of North America and the UK. However, private voice networks were separate from data networks and restrictive in their application because of the poor network technology and the monopolistic control of the PTT.

The arrival of digital networks brought a new era of communications. Digital networks offer benefits to the PTT which caused this accelerated implementation. These included:

1.   Better quality.
2.   More efficient use of the infrastructure.
3.   Lower cost.
4.   Higher speeds.
5.   The ability to manage the network efficiently.

This implementation is still occurring throughout the world-wide PTT networks, nevertheless, in many countries the bulk of the infrastructure is now based on digital connections both leased and switching.

Wide Area Network users realised the same benefits as PTTs in digital connections and in the earlier 1980s the first service became available from PTTs for digital leased services. AT&T, the deregulated long distance carrier offered T1 services to Government offices in North America.

6.3 Digital networks

The early 1980s T1 services were the start of an explosion in growth of T1 services throughout North America, with 11 million digital connections in place today. In Europe a different standard for digital connection was offered by British Telecom in the mid 1980s called 2Mbit/s or E1.

Both T1 and E1 utilise the same basic building blocks, but are constructed in a different way. The building block for T1/E1 is 64kbit/s, which is arrived at for historical reasons by the manner in which analogue voice connections are converted to a digital stream. In the 1930s, Standard Telephones and Cables Ltd. (STC), developed a technique for converting analogue voice to a digital stream called Pulse Code Modulation. This technique was adopted as standard by ITU-T as G712. The concept is as follows. The analogue network had been characterised to operate at 300Hz to 3400Hz (or 0 to 4kHz), this frequency range essentially enabling acceptable voice communication across a telecommunication infrastructure. The frequency range of 4kHz therefore requires a sample rate of twice its value in order to provide accurate reproduction.

Given this sampling rate, there was need to code the representation of the waveform. Two forms of representation were adopted, one North American called μ-law and one European called A-law. The sampling rate and the subsequent code provide an 8kbit code which represented therefore a 64kbit signal.

The foundation of digital conversion was therefore adopted, and its development was based on the earlier restrictions within the analogue network. The 64kbit building block was in turn used to form the basis of T1 (North American) and E1 (European) digital services. T1 is 24 × 64kbit/s circuits combined to form 1.536kbit/s service which is expanded by 8kbit/s to 1.544kbit/s. The 8kbit/s is used for PTT service information. (See Figure 6.6.) The signalling informa-

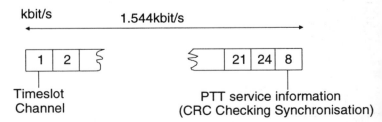

Figure 6.6   T1 transmission frame

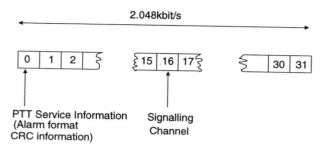

Figure 6.7 E1 transmission frame

tion is either transported as a rob bit in each channel (called timeslot) for channel associated signalling or as a complete channel (timeslot 24) for common channel signalling.

E1 on the other hand is 30 (32) × 64kbit/s combined to form 2.048kbit service. (See Figure 6.7.)

In this arrangement timeslots 1 to 15 and 17 to 31 are the 30 user usable channels. Timeslot 0 is for PTT service information, whereas timeslot 16 was preserved for signalling information either Common Channel or Channel Associated.

In due course PTTs offered lower speed digital services. In North America this was because of Digital Data Services (DDS) which offered 64kbit and below connectivity. This was followed in the late 1990s by Fractional T1 services, which simply is a full T1 delivered to the user but only a fraction of it, as desired, is utilised.

In Europe, BT was once again the first to provide 64kbit and below service under the banner of KiloStream. Now throughout Europe one is able, dependent on the PTT, to receive digital speeds below 64kbit/s, or N × 64kbit/s service (where N = 1 to 30) or 2Mbit/s (E1) or indeed Fractional E1 the equivalent to North America.

The arrival of T1 and E1 services, although initially tariffed high in the early 1980s provided the means for users to save cost; to improve quality of service; to meet the growing demands of service; and lastly to consider their network as a single point of integration for voice, data, video and image. The means to achieve these goals were originally thought to be the new modern PBX, but this was quickly

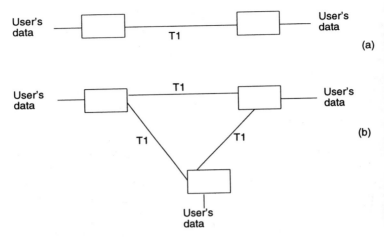

Figure 6.8   Use of multiplexers within a network: (a) point to point; (b) delta network

thrown aside by a new generation of sophisticated multiplexer, the digital data multiplexer.

It was designed by new entrants who adopted their own proprietary techniques of sending growing user data demands into T1 on E1 digital pipes. The multiplexers needed unstructured T1 or E1 pipes to capitalise on their proprietary techniques of sending data effectively through the network. Initially designed as point to point devices they effectively adapted into full networking platforms. (See Figure 6.8.)

It was not until the mid 1980s that the user realised the weakness of these 1st and 2nd generation multiplexers, whose design was principally restricted to data and proprietary in technique.

A third generation of multiplexers then appeared in the late 1980s and early 1990s. These were characterised by:

1.   The ability to integrate any form of service, be it voice, data, image, video and LAN. Connectivity would be the issue for the 1990s.

2.   Robust, evolving, flexible architecture. Adoption of new technology would be required in the 1990s.

3. Totally software driven. The ability to offer complete management capability.
4. International in design using a balance of standard and proprietary technology, i.e. Global networking.
5. A greater reach. The ability to network the smallest to the largest office from a 3 node network to a 10,000 node network.
6. Hybrid in design. The ability to operate in the PTT network or private network or both, as a single product.

Today's multiplexers adopt these basic principles and have become the main solution for wide area networking.

The features and benefits realised by such networking capabilities together with the growing reduction of digital service tariffs and their availability means their continual adoption by corporate clients into their networking connection solutions.

Today multiplexers incorporate a rich set of features to allow efficient management of the digital services.

6.3.1 Connectivity

The need to deliver a number of different services into a single integrated network requires a range of connectivity solutions, such as:

1. Voice, the ability to connect digitally from a modern PBX, but also to connect analogue devices such as E&M circuits (the original analogue private wire), telephone lines and telephone exchanges.
2. Data, the ability to handle the range of data speeds both asynchronous and synchronous from as low as 1.2kbit/s through to 1920kbit/s (and eventually to 34Mbit/s).
3. Video, the ability to handle video at rates from 64kbit/s through to 1920kbit/s.
4. Image, the ability to handle services from CAD devices with minimal delay across a network.
5. LAN, the ability to incorporate LAN interconnection devices both routers and bridges in an integrated manner.

6.3.2 Compression

Compression techniques both for voice and data offer advantages in optimising the digital service. Data compression can offer effectiveness of 8 to 1 using such techniques as statistical multiplexing.

Voice compression development however, has been more pronounced and on international circuits can realise enormous cost saving.

The compression technique adopted by the original multiplexer vendors was proprietary, called CVSD (Continuous Variable-Slope Delta Modulation).

This offers compression successfully down to 24kbit/s (nearly 3 to 1 over PCM) but is plagued by bad quality and high delays.

A technique which became standard within ITU-T G.721, called ADPCM (Adaptive Differential Pulse Code Modulation) was therefore integrated on to the earlier multiplexers and became a standard for all new generations. This offered 32kbit/s (2 to 1 compression), was of acceptable quality and minimal delay. Its quality also allowed fax to be sent across 32kbit/s unlike CVSD.

Most recently new techniques have been adapted, the most prominent being Code Excited Linear Prediction (CELP). This offers excellent quality, and with new Digital Signal Processor components, minimal delay, and will shortly be recorded as the standard for 16kbit/s voice compression by ITU-T. Using this technique 8kbit voice compression offers excellent quality and the ability to pass fax signals.

6.3.3 Network robustness

As the networks have grown so has their importance to the corporate business. To this end the multiplexer has developed techniques to protect the users' connections against failures within a network, be they the multiplexer itself or the network links. The techniques include channel re-routeing (many channels from a failed link to a working link), channel priority (nominating in order of priority the most important channels to remain in operation when limited services are available), channel down speeding (maintain a channel connection even at lower speed when resources become limited), parameter routeing (to define the manner in which channels route through a network) and resilience of both multiplexer, network management and links.

This robustness is a combination of the intelligence within a networking node and within the network management platform.

6.3.4 Family

Networks grow and change and it is therefore extremely important to support a family of product to meet different network sizes and with the inherent ability to change. To this end multiplexers are not single product categories, indeed 'the network is the product'. Within the same multiplexer family one can range from a single network connection right through $256 \times E1$ network connection, from a two multiplexer network to a 10,000 node network, and from delta (triangle) topology through to a complete mesh topology. With all these variations of network design and size, it is essential that a consistent view of the network is maintained with the same 'look and feel' no matter what the size of the multiplexer or network. Users demand a single tightly coupled view of their network not one that has a core network surrounded by alternative or even OEM access products. The reach of the network must at least have the capability to extend digitally to the smallest office in an organisation at a competitive price. This should even extend to the desk of the user, giving full management control to the network management system. Features within the multiplexer will also include:

1.   Multidrop capability to mirror analogue IBM multidrop environments.
2.   Integration into umbrella management systems such as IBM Netview, AT&T UNMA, DEC Enterprise Management Architecture, ITU-T Telephone Management Network, and of course OSI management environment.
3.   The ability to migrate to ISDN technology.
4.   Support of such capabilities as PCM bridging, and conference and virtual channel capability.

The development of digital Wide Area Networking based on the multiplexer platform is a reality of corporate private service. Its role is now either being enhanced or under threat, depending on the point of view, due to the sudden growth of Local Area Networks.

6.4 LAN Developments

6.4.1 Proliferation of personal computers

In the early 1980s Personal Computers (PC) infiltrated the office environment on a large scale. PCs were purchased either as replacements for larger, and subsequently out-dated systems or as part of an office automation program.

The replacement of older, (i.e., mainframe/mini) systems was the result of innovative technology that placed tremendous processing power in a small desktop device. Users who required pure computing power, but not necessarily access to remote information, benefited directly from this migration to PCs.

PCs were also purchased in an effort to improve productivity by automating manual processes, such as word processing, filing and accounting.

At this point in time, the vast majority of PCs were installed as stand alone devices. That is to say, they were not attached to any other devices, except perhaps a printer. External communications, such as to other PCs within the same work area did not exist. However, as the computing power of the PC increased, and the need to make use of, and share the information on PCs also increased connectivity, or networking, became a necessity.

6.4.2 Evolution of Local Area Networks

The requirement for computer connectivity led to the development of the modern day Local Area Network (LAN).

The first LANs to appear on the market in the early 1980s were represented by many different proprietary networking schemes from a multitude of vendors.

Most of these products featured low speed, (under 1 Mbit/s), bus or serial connections between computers, with little if any software control of the network. As such, these early networks were complicated and difficult to use.

The market today is very different. Vendors such as Novell, Banyan and Microsoft have produced software-driven network operating systems that can function over a multitude of standards-based LANs. Standards such as IEEE 802.3 (Ethernet), and IEEE 802.5 (Token Ring), have provided reliable and functional methods for interconnecting PCs. In addition, many LANs also connect mainframe and mini computer systems.

6.4.3 LAN applications

Local Area Networks can be utilised to provide any or all of the following facilities (see Figure 6.9):

1. The interconnection of all computers, terminals, PCs and other workstations within a department or building.
2. The effective communication between all devices attached to the network, regardless of their vendor.

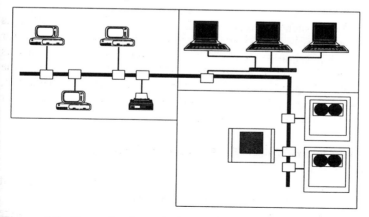

Figure 6.9   Example of a typical LAN application

3. The sharing of expensive peripheral resources attached to the network such as disks, printers and plotters, central processors, and databases.
4. The standardisation of wiring, hardware and software required for communication and applications.
5. The establishment of a foundation for continued growth and the expansion of a distributed computing environment.

6.4.4 LAN characteristics

The key characteristics of a LAN are:

1. Transmission medium.
2. Signalling technique.
3. Access method.
4. Network topology.

6.4.4.1 *Transmission media*

The transmission medium is the physical entity used to convey data over the LAN and includes any active hardware, such as amplifiers, required to regenerate the signal on the medium, and any passive hardware, such as taps and connectors, required to provide access to the medium.

The most common types of installed media are:

1. Broadband coaxial cable.
2. Baseband coaxial cable.
3. Twisted pair wire.

In addition, optical fibre is emerging as an increasingly important transmission medium.

6.4.4.2 *Signalling techniques*

LAN signalling techniques can be divided into baseband and broadband transmission.

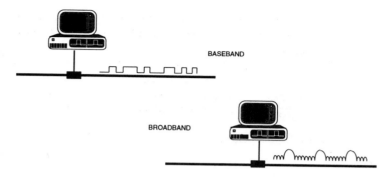

Figure 6.10   Signalling techniques on LANs

Baseband transmission utilises direct encoding to convey digital information in its digital form, to the transmission medium. In baseband transmission, only one channel is available to all network users, therefore only one signal can survive on the transmission medium at any one time.

Broadband transmission utilises radio frequency modems to convey data signals over the transmission medium. Because the available bandwidth is divided into separate channels, the use of a frequency division multiplexing (FDM) allows several 'conversations' to coexist on the LAN simultaneously.

Baseband LANs are generally less expensive and easier to install than broadband LANs. (See Figure 6.10.)

6.4.4.3 *Access methods*

The method of access to the LAN permits all the devices attached to the transmission medium to share that medium in a controlled fashion.

The two most common access methods are Carrier Sense Multiple Access with Collision Detection (CSMA/CD) and Token Passing. (See Figure 6.11.)

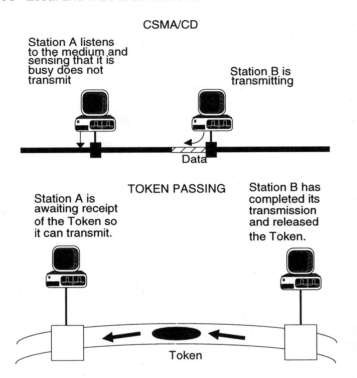

Figure 6.11   LAN access methods

CSMA/CD permits any device attached to the medium to transmit on that medium (multiple access), if it senses that the medium is free (carrier sense). Occasionally two or more devices simultaneously sense that the medium is free and begin to transmit. This creates a collision which is detected by the sending device and the data is retransmitted after a random time.

Token Passing permits each device on the LAN to control the medium for a predetermined maximum time while it is in possession of a message packet (token). The token is passed from workstation to workstation on the LAN until the one wishing to transmit receives it.

When the workstation has finished transmitting, the token is passed to the next device in a predetermined sequence.

6.4.4.4 *Network topology*

Several network topologies, (physical layout), can be used to design a Local Area Network, the most common being Bus, Ring, Star and Star-Shaped Ring. (See Figure 6.12.):

1. Bus network is the most prevalent LAN configuration, used in both Ethernet (CSMA/CD) and Token Passing Bus. All devices connect to the bus by means of a tap, (transceiver).
2. In Ring topologies, all networked devices are connected in a closed loop.
3. For Star topologies the networked devices are connected directly to a central network device, which is usually active.
4. Star-Shaped Ring networks use dedicated cabling to each workstation, which converges together in a wiring closet.

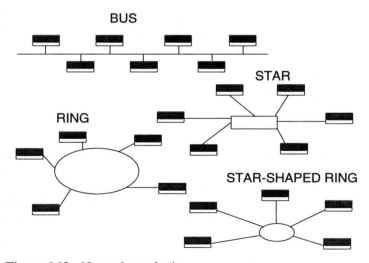

Figure 6.12 Network topologies

6.4.4.5 *LAN standards*

Because the first LAN systems were based on vendor specific technology, customers were forced to deal with single suppliers. In order to allow users the freedom to purchase and interconnect equipment from multiple vendors, it was clear that equipment standardisation was required.

The two most influential bodies in determining LAN standards are the Institute of Electrical and Electronic Engineers, (IEEE), and the International Standards Organisation, (ISO).

The IEEE Local Network Standards Committee, (Project 802), is developing LAN access standards and protocols in a layered approach similar to the ISO's Open Systems Interconnection (OSI) Reference Model.

The OSI Reference Model (see Chapter 1) represents the relationship between a network and the services it can support by a hierarchy of seven protocol layers. Each layer uses the services of the lower layers in conjunction with its own functions to create new services which are then made available to the higher layers. (See Figure 6.13.)

Sub groups within the IEEE 802 Committee are producing the following LAN standards:

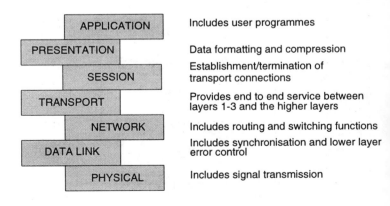

APPLICATION	Includes user programmes
PRESENTATION	Data formatting and compression
SESSION	Establishment/termination of transport connections
TRANSPORT	Provides end to end service between layers 1-3 and the higher layers
NETWORK	Includes routing and switching functions
DATA LINK	Includes synchronisation and lower layer error control
PHYSICAL	Includes signal transmission

Figure 6.13 The OSI reference model

1. IEEE 802.1, covering architecture, addressing, internetworking and management.
2. IEEE 802.2, Logical Link Control (LLC) Protocol, common to the various types of media implementation.
3. IEEE 802.3, Carrier Sense Multiple Access and Collision Detect.
4. IEEE 802.4, Token Passing Bus.
5. IEEE 802.5, Token Passing Ring.
6. IEEE 802.6, Metropolitan Networks.

The ISO has adopted the IEEE standards as ISO 8802 and relates the IEEE model to its own. (See Figure 6.14.)

The IEEE 802 series are standards covering access methods and as such are applicable over several types of media, such as coaxial cable, twisted pair wiring and optical fibre. The IEEE has developed a classification system which groups local area networks according to speed, signalling technique and segment length.

For example, the most popular type of 802.3/Ethernet LAN is 10 Base 5. This refers to LANs over the common, thick, yellow coaxial

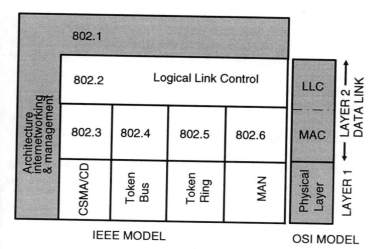

Figure 6.14 IEEE and OSI models

cable, (YE-50). The speed at which data can theoretically travel over this medium is 10Mbit/s, the signalling technique is baseband and the maximum segment length without signal regeneration is 500 metres, hence 10 Base 5. This particular standard requires the special 'thick' Ethernet cable, most often yellow or bright orange in colour, with multiple layers of shielding. Other 802.3 LAN standards include:

1.   10 Base 2, also known as Thinnet or Cheapernet.
2.   1 Base 5, the original standard for the AT&T StarLan[TM]
3.   10 Base T, applicable to LANs over twisted pair wiring.
4.   10 Base F, applicable to LANs over an optical fibre.
5.   10 Broad 36, CSMA/CD using broadband signalling techniques.

IEEE 802.5 has standards governing nine types of physical media, including shielded and unshielded twisted pair at 1, 4, and 16Mbit/s.

Table 6.1 shows the characteristics of some popular LANs.

6.5 Internetworking devices

Although local area networks have been in existence in one form or another for many years, the unprecedented growth of LANs during the latter half of the 1980s made network users increasingly aware of the limitations of their installed networks, and refocused attention on ways to overcome them.

The first problems realised were those concerned with the physical distance of the network. A typical 10 Base 5 coaxial Ethernet, for example, has a maximum segment length of 500 metres and may only accommodate up to 100 'taps' per segment.

The requirement to extend the length of the network led to the development of what is commonly acknowledged as the first internetwork device, the repeater.

6.5.1 The repeater

The repeater is the simplest form of internetworking device and evolved first in conjunction with CSMA/CD-type networks in the

Tab;e 6.1   Characteristics of some popular LANs

Signall-ing tech-nique	Access method			Trans-mission medium	Network topology
	802.3	802.4	803		
Broad-band	Yes	Yes	No	Coax	Bus
	Yes	Yes	No	Coax	Bus/Tree
Baseband	Yes	No	No	Coax	Bus
	Yes	No	No	Twisted Pair	Bus
	Yes	No	No	Multi-Wire	Bus
	Yes	No	No	Coax	Bus/Star
	Yes	No	No	Fibre Optic	Bus/Star
	Yes	No	No	Coax	Star
	Yes	No	No	Twisted Pair	Star
	No	No	Yes	Coax	Ring
	No	No	Yes	Twisted Pair	Ring
	No	No	Yes	Fibre	Ring

mid 1970s. It operates at the lowest layer of the ISO Reference Model, i.e., the Physical Layer. (See Figure 6.15.)

Repeaters can extend the physical range of a LAN by collecting the stream of electrical impulses on one LAN and repeating that signal on to another identical LAN. Repeaters, however are limited in their ability to extend LANs. This is because LAN architectures incorporate certain assumptions regarding propagation delays, and as repeaters are added to networks, propagation delays eventually in-

Figure 6.15   OSI model of a repeater

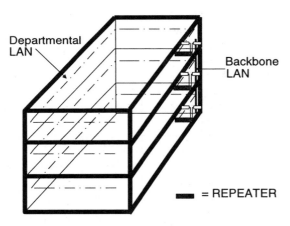

Figure 6.16   Repeater application within a building

crease beyond tolerable thresholds. In addition, use of repeaters often requires careful and detailed network design and planning. (See Figure 6.16.)

Since they operate at the Physical Layer, repeaters perform independently of the higher-level processing required in more complex networks. Therefore, they can effectively operate at the same speeds as the extended network. This, however, also means that repeaters are only capable of interconnecting LANs with similar protocol formats.

A further advantage to the deployment of repeaters in extending a LAN is that they can effectively isolate cable faults to a single segment of the extended LAN. On the other hand, indiscriminate use of repeaters can create problems with network performance and station-to-station accessibility.

6.5.2 Bridges

As more and more devices become attached to the LAN, the utilisation increases. As the amount of traffic on the network approaches the prescribed operating limits of the LAN, network efficiency begins to decrease.

For example, on a non-contention LAN, such as an IEEE 802.5 Token Passing Ring, the station must wait a finite time before the token makes its way around the ring to it, thereby enabling it to transmit. This minimum time to transmission increases with the number of attached devices. Similarly, on a contention LAN, such as an IEEE 802.3 CSMA/CD, as the number of devices contending for control of the medium increases, the number of collisions on the network also rises, thereby reducing its overall performance.

Analysis of the nature of traffic on the LAN, in most cases, led to the conclusion that effectively segmenting the network into two or more separate LANs could increase overall network performance.

This would certainly be the case when the network could be segmented into heavy users, (e.g. CAD/CAM users, computer rooms), and light users, (e.g. word processing). (See Figure 6.17.)

A further advantage of such segmentation is that failures on a particular segment of the LAN can be contained within that segment.

The solution to the congestion problem led to the development of the bridge, or to be more precise, the local bridge.

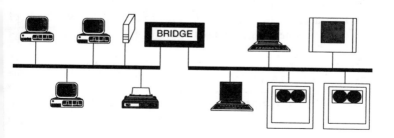

Figure 6.17 LAN interconnection using a local bridge

Since a local bridge could be used to create two networks out of a single network, it followed that it could also be employed to create one logical extended network out of two, (or more), locally dispersed LANs.

6.5.2.1 *Local bridges*

Bridges operate at the Data Link Layer, (Layer 2), of the OSI Model, or more specifically at the Media Access Control, (MAC), sublayer. (See Figure 6.18.)

Upon receipt of a data packet, bridges examine the source and destination address of the data packet. If the destination device is on a network other than that of the source device, then the bridge will 'FORWARD' the packet onto the extended network. If the destination device address is on the same network segment as the source device, then the bridge will not forward the packet; instead it will block its path onto the extended network effectively keeping it local. In this way the bridge acts as a 'FILTER' of data packets.

By means of 'Filtering and Forwarding' bridges can create one single logically unified network out of several locally discrete LANs, while at the same time limiting the flow of unnecessary traffic between them.

Organisations that had local area networks within their geographically dispersed operating sites soon began to express a desire to interconnect all their remote LANs together, to effectively build a single organisation-wide area network, irrespective of geography.

Manufacturers responded to this requirement with two types of solutions, remote bridges and routers.

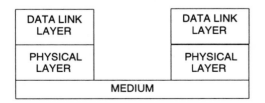

Figure 6.18   OSI model of a bridge

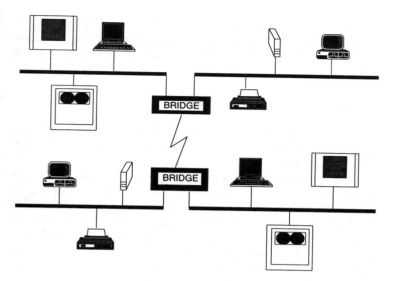

Figure 6.19 LAN interconnection using remote bridges

Manufacturers offered as a simple, 'transparent' solution, bridges that incorporated wide area interfaces to provide access to different types of media, such as leased or switched analog telephone lines, 48, 56, 64kbit/s lines, or E1 or T1 lines.

This led to the development of the remote bridge which performed exactly the same functions as a local bridge, with the exception that it could now be used to build a single logical network, unrestricted by geography. Since this type of bridge had wide area interfaces the user typically had to have two identical bridges at either end of the communications link. Thus, remote bridges were often referred to as 'half bridges'. (See Figure 6.19.)

6.5.3 Routers

The second solution to the problem of interconnecting LANs over unlimited geographical distances was the router.

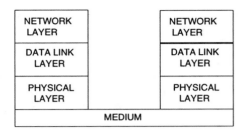

Figure 6.20 OSI model of a router

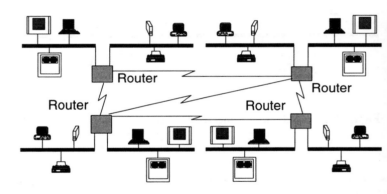

Figure 6.21 Example of an extended network using routers

This device operates at the Network Layer of the OSI Model, (Layer 3). Routers essentially offer selective routeing of individual packets over multiple communications paths. (See Figure 6.20.)

Routers have the capability to transmit data packets over different paths in an extended network depending on the network user's priorities, such as the least costly, fastest, or most direct route.

Because they operate at a higher layer in the OSI Model than bridges, routers have a greater level of 'understanding' of the data passing on the LAN. This allows routers to create a logically extended network comprised of separate subnetworks, in contrast to the single logically unified network constructed by bridges. (See Figure 6.21.)

The extra processing required by routers to manipulate data packets, however, impacts upon throughput performance. Typically, conventional routers introduce longer delays in getting packets from source node to destination node, resulting in slower response times, or require more processing power to provide the same response times. As Network Layer connectors, routers are protocol specific; indeed the earliest routers could only connect LANs with identical Network Layer protocols.

6.5.4 Gateways

Network users who want to connect two totally dissimilar LANs together use a 'gateway'. There is still some confusion regarding the use of this term. Many people still refer to routers as gateways. The following definitions are now more widely accepted:

1. Router: a device that operates at Layer 3 of the OSI Model.
2. Gateway: a device that operates at Layers 4, 5, 6 and 7, (i.e. the Transport, Session, Presentation and Application Layers) of the OSI Model.

Gateways effectively provide a protocol conversion service, at each of the higher layers of the OSI Model for the interconnection of LANs. Therefore, LANs with separate protocol formats such as NetBios and SNA can communicate through a gateway.

Similarly, gateways permit different machine types to communicate. For example, IBM PCs and asynchronous terminals can communicate with IBM hosts through gateways via protocol emulation.

Gateways operating at the Session, Presentation and Application Layers allow diverse architectures such as SNA or DECnet to communicate. Such gateways may make use of common 'lower layer' network facilities such as X.25 packet networks. (See Figure 6.22.) Operating at the Application Layer, gateways can enable specific applications, such as different electronic mail systems to communicate.

Gateways were primarily designed for specific environment-to-environment connectivity and not particularly for the high speeds

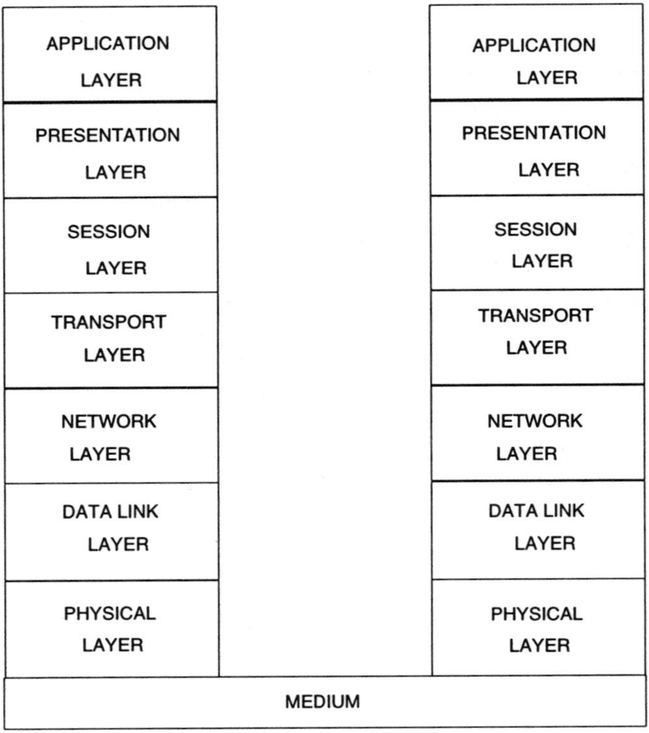

APPLICATION LAYER		APPLICATION LAYER
PRESENTATION LAYER		PRESENTATION LAYER
SESSION LAYER		SESSION LAYER
TRANSPORT LAYER		TRANSPORT LAYER
NETWORK LAYER		NETWORK LAYER
DATA LINK LAYER		DATA LINK LAYER
PHYSICAL LAYER		PHYSICAL LAYER
MEDIUM		

Figure 6.22 OSI model of a gateway

common to LAN communications. Since they operate at the highest layers of the OSI Model, gateways are required to perform substantially more data processing than the other internetworking devices. They are, therefore, slower, less transparent, less flexible and perhaps the most network specific of all the LAN interconnectivity devices. (See Figure 6.23.)

Because both repeaters and gateways are network specific, bridges and routers emerged as the most popular solutions to internetworking problems.

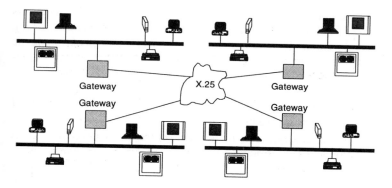

Figure 6.23 Example of an extended network using gateways

Nevertheless, considerable confusion remains as to when it is best to bridge and when it is best to route. This confusion has been compounded by several factors. Firstly, conventional bridge functionality has expanded into the territory originally monopolised by routers. Secondly, routers are now performing at speeds traditionally associated with bridges, and thirdly the developments in both bridge and router technology have been accompanied by a host of totally new hybrid internetworking products.

6.5.5 Developments in bridging technology

The first major development in bridging was the addition of 'learning'. Learning enabled bridges to automatically track the source addresses of data packets received by the bridge in order to build up a learned address table of locally attached devices. This capability enabled the bridge to automatically and dynamically filter local segment traffic from the rest of the network.

A network management capability has been added on more intelligent bridging devices to provide extended network management performance statistics and remote configuration of bridges.

One of the problems of implementing bridges in complex network topologies, particularly networks that involve looped data paths, is the likelihood of a 'broadcast storm'. That is, certain types of packets could be perpetually circulated around the extended network causing a degradation of throughput performance and eventually bringing the network down. To overcome this problem, many bridges now implement the IEEE 802.1(D) Spanning Tree algorithm.

The Spanning Tree algorithm decides which bridge in the network is to become the 'root bridge', based upon the number and speed of the paths between the bridges.

The bridge deemed to be the root bridge is allocated the highest priority of all the bridges in the network and the network is then configured and routes determined.

Effectively, the bridges automatically disable selected routes within the extended network to create a 'loop-free' network. This in effect changes a looped or 'mesh' network into a tree structure for which there is one and only one active path from one station to any and every other station. However, the Spanning Tree algorithm carries the disadvantage that alternate paths, and in many cases, expensive wide area network bandwidth, remain unused and become active only when other paths fail.

The latest bridges are now capable of automatically configuring themselves upon insertion into a network, and have become true 'plug and play' devices.

6.5.6 Developments in router technology

While developments in bridge technology were occurring, router technology was also advancing.

Conventional routers only supported the interconnection of local area networks that had identical characteristics (e.g. Ethernet networks). Similarly, these routers only operated for a single routeing protocol over the extended network, such as TCP/IP or DECnet.

The latest generation of routers, however, now supports a variety of LAN types and concurrently supports more than one routeing protocol.

The speeds at which routers process data have also increased, and in some cases performance now almost matches that of a straightforward bridging device.

6.5.7 Hybrid interconnection devices

The concurrent developments in bridging and routeing technology led to the emergence of hybrid interconnection products such as the router bridge, sometimes referred to as bridge/routers, or brouters.

These devices are capable of operating at both layers 2 and 3, (Data Link and Network respectively), of the OSI model.

A router/bridge can provide a Network Layer (routeing) service to one or more protocols that it has been programmed to support, while at the same time offering a Data Link Layer (bridging) service to protocols that it does not recognise. Therefore, a router/bridge can simultaneously route the DECnet protocol and bridge the DEC LAT protocol, (which cannot be routed) over the same link.

The term brouter is also commonly applied to a bridge (Layer 2) device, which rather than using the standards based Spanning Tree algorithm, uses a proprietary scheme for avoiding the problems introduced by mesh or 'looped' topologies, while at the same time taking advantage of multiple and alternate paths. (See Figure 6.24.)

6.6 LAN/WAN challenge for the 1990s

Following the wide scale adoption of LANs in the 1980s, bridges became and remain, popular due mainly to their simplicity, ease of installation, and low cost. A bridging solution performs well in small-to-medium-sized departmental environments.

However, as networking continues to evolve, so does the requirement for effective management of the extended network. Bridges are only capable of building one logically extended network, unlike routers which can establish independently managed networks, (or sub-networks), within the extended network.

Router/bridges attempt to provide the best of both technologies, offering the simplicity and performance of bridging and the network control capabilities of routeing, to help resolve the ever increasing complexities of LAN-to-LAN communications.

In the last two sections we covered the development of two discrete network technologies, WAN and LAN.

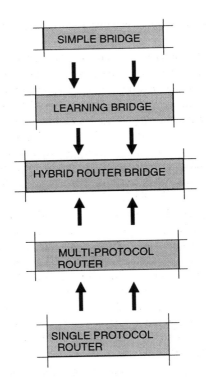

Figure 6.24 Evolution of the hybrid router bridge

The Wide Area Network, initially based on analogue modem connectivity, developed to a sophisticated digital networking infrastructure. Local Area Networks have rapidly been introduced in the networking environment, first as discrete networks and then as interconnection into the analogue or digital infrastructure.

The 1990s and into the next century are set to see an expansion of both technologies, both as separate networking solutions, but most importantly as single network solution under the apt name 'All Area Networking' (AAN).

The traditional wide area supplier has quickly realised the importance of integrating LAN interconnect as a unified product within his overall product line. Unified product means not only the capability to be managed under the same network platform, but to operate with the same physical attributes as all his existing connectivity elements.

The rising local area interconnection suppliers have attempted to brush aside the issue relating to other forms of networking, but concentrate on improving the LAN capability. This under the belief that all private networking will be dominated by the needs of LANs rather than voice, video, traditional data service and image. This is not an assumption but a realisation that if tariffed correctly all these other services can be either replaced by LANs or in the case of voice or video replaced by the existing but improved public switched network (be that in an ISDN form).

In some respects this was a real threat, since it was clear that routeing LAN traffic via a multiplexer has no real benefits. This was true until the emergence of a new standard called Frame Relay. Frame Relay see Chapter 4) which is based on ITU-T I.120 standards provides an effective method of handling multiple LANs into a single or multisystem wide area network. As has been the case previously, the most reasonable platform for this type of network would be a multiplexer. Frame Relay efficiency handles the packet type connectivity of LAN by optimising the infrastructure to meet the dramatic demand for LAN traffic.

A network of the very near future will combine all the traditional attributes of digital multiplexers together with the ability to handle LAN interconnection, either as part of the multiplexer or as a separate access device on the network, but all under a single management environment.

Yet the challenge of the 1990s is not yet over with the arrival of new standard transmission techniques based on Synchronous Digital Hierarchy (SDH) or as it is called in North America SONET. These allow transmission planners to construct networks on single building blocks, with higher bandwidth using fibre error free connection, as a standard product from many vendors and most importantly the ability to integrate into the existing inefficient and proprietary networks they have developed. This is followed by a new network switching infrastructure called (see Chapter 5) Asynchronous Transport Mode (ATM), which caters for the growing demand of higher switching connectivity and packet mode of both LAN, voice, video and image. Also included is the arrival of management architectures based on OSI standard and with the ability to separate/partition services and function from this single management architecture.

The changes within the network during 1990s and early next century will see an ever demanding role for WAN and LAN networking, which will ultimately be under the same networking banner AAN, all working to the goal 'the network is the product'.

6.7 Acknowledgement

Acknowledgement to Steve Gerrard, Marketing Manager for local and wide area integration, with Newbridge Networks, who practically wrote the LAN section.

6.8 References

Bell, M. (1995) Going farther afield, *LAN Magazine*, June.

Biesaart, R. (1994) LAN interconnection: ISDN in its own right, *Telecommunications*, October.

Broadhead, S. (1993) The path to gateways, *Datacom*, September.

Cobbe, T. (1993) MAN strategies become a commercial reality, *Telecommunications*, May.

Communicate (1993) Where are gateways going? *Communicate*, February.

Communicate (1994) Ethernet still switches users on, *Communicate*, September.

Guidoux, L. (1995) Intelligent solutions for data communications networks, *Telecommunications*, June.

Hunter, P. (1994) To bridge or not to bridge? *Communications News*, June.

Joyce, R. (1994) Hubs: a stacked future, *Telecommunications*, June.

Nye, R. (1990) A LAN guide for the perplexed, *Connexion*, 18 July.

Roman, R. (1992) Network approaches for third generation hubs, *Inte Net*, July.

Smith, P. (1994) LAN interworking — connectivity or networking? *Telecommunications*, May.

Valiani, M. (1993) The fading distinction between LANs and WANs, *Telecommunications*, April.

Valovic, T. (1989) Metropolitan Area Networks: a status report, *Telecommunications*, July.

Wilson, R. (1993) Nice and easy does it ..., *Electronics Weekly*, 17 November.

Zitsen, W. (1990) Metropolitan Area Networks: taking LANs into the public network, *Telecommunications*, June.

7. Acronyms

Every discipline has its own 'language' and this is especially true of telecommunications, where acronyms abound. In this guide to acronyms, where the letters within an acronym can have slightly different interpretations, these are given within the same entry. If the acronym stands for completely different terms then these are listed separately.

AAN	All Area Networking. (Networking covering local and wide areas. Also used to imply combined use of LAN and WAN.)
ACK	Acknowledgement. (Control code sent from a receiver to a transmitter to acknowledge the receipt of a transmission.)
ACSE	Association Control Service Element. (OSI application.)
ADM	Add-Drop Multiplexer. (Term sometimes used to describe a drop and insert multiplexer.)
ADP	Automatic Data Processing.
ADPCM	Adaptive Differential Pulse Code Modulation. (ITU-T standard for the conversion and transmission of analogue signals at 32kbit/s.)
AM	Amplitude Modulation. (Analogue signal transmission encoding technique.)
ANDF	Architecture Neutral Distribution Format. (Scheme from OSF to enable software to be produced in single code to run on any hardware.)
ANSI	American National Standards Institute.
ANI	Automatic Number Identification. (Feature for automatically determining the identity of the caller.)
APC	Adaptive Predictive Coding.
API	Application Programming Interface.

APK   Amplitude Phase Keying. (A digital modulation technique in which the amplitude and phase of the carrier are varied.)

ARQ   Automatic Request for repetition. (A feature in transmission systems in which the receiver automatically asks the sender to retransmit a block of information, usually because there is an error in the earlier transmission.)

ASCII   American Standard Code for Information Interchange. (Popular character code used for data communications and processing. Consists of seven bits, or eight bits with a parity bit added.)

ASN.1   Abstract Syntax Notation One.

ASR   Automatic Send/Receive. (Operation usually carried out by an older type of teleprinter equipment.)

ASK   Amplitude Shift Keying. (Digital modulation technique.)

ATDM   Asynchronous Time Division Multiplexing.

ATM   Asynchronous Transfer Mode. (ITU-T protocol for the transmission of voice, data and video.)

AU   Administrative Unit. (Term used in synchronous transmission. Ref: ITU-T C.709. It is the level at which circuit administration is carried out by the operator.)

AVDM   Analogue Variable Delta Modulation.

B-ISDN   Broadband Integrated Services Digital Network.

BABT   British Approvals Board for Telecommunications.

BASIC   Beginners All Symbolic Instruction Code. (Computer programming language.)

BCC   Block Check Character. (A control character which is added to a block of transmitted data, used in checking for errors.)

BCD   Binary Coded Decimal. (An older character code set, in which numbers are represented by a four bit sequence.)

BCH	Bose Chaudhure Hocquengherm. (Coding technique.)
BELLCORE	Bell Communications Research. (Research organisation, incorporating parts of the former Bell Laboratories, established after the divestiture of AT&T. Funded by the BOCs and RBOCs to formulate telecommunication standards.)
BER	Bit Error Ratio. (Also called Bit Error Rate. It is a measure of transmission quality. It is the number of bits received in error during a transmission, divided by the total number of bits transmitted in a specific interval.)
BERT	Bit Error Ratio Tester. (Equipment used for digital transmission testing.)
BIP	Bit Interleaved Parity. (A simple method of parity checking.)
BISYNC	Binary Synchronous communications. (Older protocol used for character oriented transmission on half-duplex links.)
BnZS	Bipolar with n-Zero Substitution. (A channel code. Examples are B3ZS which has three-zero substitution; B6ZS with six-zero substitution, etc.)
BOC	Bell Operating Company. (Twenty-two BOCs were formed after the divestiture of AT&T, acting as local telephone companies in the US. They are now organised into seven Regional Bell Operating Companies or RBOCs.)
BORSCHT	Battery, Overload protection, Ringing, Supervision, Coding, Hybrid, and Test access. (These are the functions provided in connection with a subscriber line circuit. The functions are usually implemented by an integrated circuit.)
BPON	Broadband over Passive Optical Networks.
BPSK	Binary Phase Shift Keying.
BPV	Bipolar Violation. (Impairment of digital transmission system, using bipolar coding, where two pulses occur consecutively with the same polarity.)

BRAP	Broadcast Recognition with Alternating Priorities. (Multiple access technique.)
BRZ	Bipolar Return to Zero. (A channel coding technique, used for digital transmission.)
BSI	British Standards Institute.
BUNI	Broadband User Network Interface.

CAD/CAM	Computer Aided Design/Computer Aided Manufacture.
CAS	Channel Associated Signalling. (ITU-T signalling method.)
CASE	Computer Aided Software Engineering (or Computer Aided System Engineering).
CBEMA	Computer Business Equipment Manufacturers' Association. (USA.)
CCD	Charged Coupled Device. (Semiconductor device used for analogue storage and imaging.)
CCIA	Computer and Communications Industry Association. (USA.)
CCIR	Comite Consultatif Internationale des Radiocommunications. (International Radio Consultative Committee. Former standards making body within the ITU and now part of its new Radiocommunication Sector.)
CCIS	Common Channel Interoffice Signalling. (North American signalling system which uses a separate signalling network between switches.)
CCITT	Comite Consultatif Internationale de Telephonique et Telegraphique. (Consulative Committee for International Telephone and Telegraphy. Standards making body within the ITU, now forming part of the new Standardisation Sector.)
CCR	Commitment, Concurrency and Recovery.
CCS	Common Channel Signalling. (ITU-T standard singalling system. Also called CCSS.)

CCSS Common Channel Signalling System. (ITU-T standard signalling system. Also called CCS or Number 7 signalling.)

CDM Code Delta Modulation. (Or Continuous Delta Modulation.)

CDMA Code Division Multiple Access.

CDPD Cellular Digital Packet Data. (Data transmission technique using cellular voice networks.)

CEC Commission of the European Communities.

CEE Central and Eastern Europe.

CEN Comite Europeen de Normalisation. (Committee for European Standardisation.)

CENELEC Comite Europeen de Normalisation Electrotechnique. (Committee for European Electrotechnical Standardisation.)

CEPT Conference des administrations Europeenes des Postes et Telecommunications. (Conference of European Posts and Telecommunications administrations. Body representing European PTTs.)

CFM Companded Frequency Modulation.

CFSK Coherent Frequency Shift Keying.

CLAN Cableless Local Area Network. (Radio based LAN.)

CLIP Connection-Less Interworking Protocol. (OSI Network Layer.)

CLTS Connection-Less Transport Service.

CMI Code Mark Inversion. (Line coding technique.)

CMIP Common Management Information Protocol. (Protocol widely used in network management.)

CMIS Common Management Information Service.

CMISE Common Management Information Service Element. (Specific type of ASE.)

CMOL Common Management information protocol Over Logical link control.

CMT Character Mode Terminal. (e.g. VT100, which does not provide graphical capability.)

CODEC COder-DECoder.

COS	Corporation for Open Systems. (US trade association.)
COSINE	Co-operation for Open Systems Interconnection Networking in Europe.
CPE	Customer Premise Equipment.
CPFSK	Continuous Phase Frequency Shift Keying.
CPODA	Contention Priority-Oriented Demand Assignment protocol. (Multiple access technique with contention for reservations. See PODA and FPODA.)
CPSK	Coherent Phase Shift Keying.
CPU	Central Processing Unit. (Usually part of a computer.)
CR	Carriage Return. (Code used on a teleprinter to start a new line.)
CRC	Cyclic Redundancy Check. (Bit oriented protocol used for checking for errors in transmitted data.)
CSDN	Circuit Switched Data Network.
CSMA	Carrier Sense Multiple Access. (LAN Multiple access technique.)
CSMA/CD	Carrier Sense Multiple Access with Collision Detection. (LAN access technique, with improved throughput, under heavy load conditions, compared to pure CSMA.)
CVSD	Continuous Variable Slope Delta modulation. (Proprietary method used for speech compression. Also called CVSDM.)
DAMA	Demand Assigned Multiple Access.
DARPA	Defence Advanced Research Projects Agency. (USA Government agency.)
DASS	Digital Access Signalling System. (Signalling system introduced in the UK prior to ITU-T standards I.440 and I.450.)
DASS	Demand Assignment Signalling and Switching unit.
DCC	Data Communication Channel.

DCDM	Digitally Coded Delta Modulation. (Delta modulation technique in which the step size is controlled by the bit sequence produced by the sampling and quantisation.)
DCE	Data Circuit termination Equipment. (Exchange end of a network, connecting to a DTE. Usually used in packet switched networks.)
DCF	Data Communications Function.
DCN	Data Communications Network.
DCR	Dynamically Controlled Routing. (Traffic routing method proposed by Bell Northern Research, Canada.)
DDI	Distributed Data Interface. (Proposal to run the FDDI standard over unshielded twisted pair.)
DDN	Digital Data Network.
DDN	Defence Data Network. (US military network, derived from the ARPANET.)
DDP	Distributed Data Processing.
DDS	Digital Data Service. (North American data service.)
DE	Defect Events. (e.g. loss of signal, loss of frame synchronisation, etc. ITU-T M.550 for digital circuit testing.)
DEDM	Dolby Enhanced Delta Modulation.
DEPSK	Differentially Encoded Phase Shift Keying.
DES	Data Encryption Standard. (Public standard encryption system from the American National Bureau of Standards.)
DFT	Discrete Fourier Transform.
DLC	Data Link Control.
DLS	Data Link Service.
DM	Delta Modulation. (Digital signal modulation technique.)
DMA	Direct Memory Access.
DME	Distributed Management Environment. (OSF.)
DNIC	Data Network Identification Code. (Part of an international telephone number.)

DOD	Department Of Defence. (US agency.)
DOV	Data Over Voice. (Technique for simultaneous transmission of voice and data over telephone lines. This is a less sophisticated technique than ISDN.)
DPCM	Differential Pulse Code Modulation.
DPNSS	Digital Private Network Signalling System. (Inter-PABX signalling system used in the UK.)
DQDB	Distributed Queue Double Bus. (IEEE standard 802.6 for Metropolitan Area Networks.)
DQPSK	Differential Quaternary Phase Shift Keying.
DRG	Direction a la Reglementation Generale. (Directorate for General Regulation, in France.)
DS-0	Digital Signal level 0. (Part of the US transmission hierarchy, transmitting at 64kbit/s. DS-1 transmits at 1.544Mbit/s, DS-2 at 6.312Mbit/s, etc.)
DSAP	Destination Service Access Point. (Refers to the address of service at destination.)
DSE	Data Switching Exchange. (Part of packet switched network.)
DSI	Digital Speech Interpolation. (Method used in digital speech transmission where the channel is activated only when speech is present.)
DSM	Delta Sigma Modulation. (Digital signal modulation technique.)
DSP	Digital Signal Processing.
DSS	Digital Subscriber Signalling. (CCIT term for the N-ISDN access protocol.)
DSSS	Direct Sequence Spread Spectrum.
DSU	Data Service Unit. (Customer premise interface to a digital line provided by a PTT.)
DTE	Data Terminal Equipment. (User end of network which connects to a DCE. Usually used in packet switched networks.)
DTI	Department of Trade and Industry.
DTMF	Dual Tone Multi-Frequency. (Telephone signalling system used with push button telephones.)

EBCDIC	Extended Binary Coded Decimal Interchange Code. (Eight bit character code set.)
EBIT	European Broadband Interconnect Trial. (Collaborative effort between PTOs to support RACE application pilots into switched broadband. Initial trial was planned at 2Mbit/s switched, progressing to 140Mbit/s.)
EBU	European Broadcasting Union.
EC	European Commission.
ECC	Embedded Communication Channel. (Channel used within SDH to carry communication information rather than data.)
ECC	Error Control Coding. (Coding used to reduce errors in transmission.)
ECMA	European Computer Manufacturers Association.
ECSA	Exchange Carriers Standards Association. (USA)
ECTRA	European Committee for Telecommunications Regulatory Affairs. (Part of CEPT.)
EEC	European Economic Community.
EIA	Electronic Industries Association. (Trade association in USA)
EOA	End Of Address. (Header code used in a transmitted frame.)
EOB	End Of Block. (Character used at end of a transmitted frame. Also referred to as End of Transmitted Block or ETB.)
EOC	Embedded Operations Channel. (Bits carried in a transmission frame which contain auxiliary information such as for maintenance and supervisory. This is also called a Facilities Data Link, FDL.)
EOT	End Of Transmission. (Control code used in transmission to signal the receiver that all the information has been sent.)
EPHOS	European Procurement Handbook for Open Systems. (Equivalent to GOSIP.)
ESF	Extended Superframe. (North American 24 frame digital transmission format.)

ETB	End of Transmission Block. (A control character which denotes the end of a block of Bisync transmitted data.)
ETE	Exchange Terminating Equipment.
ETNO	European Telecommunications Network Operators. (Association of European public operators.)
ETS	European Telecommunication Standard. (Norme Europeenne de Telecommunications. Standard produced by ETSI.)
ETSI	European Telecommunications Standards Institute.
ETX	End of Text. (A control character used to denote the end of transmitted text, which was started by a STX character.)
EUCATEL	European Conference of Associations of Telecommunication industries.)
EWOS	European Workshop on Open Systems.

FAS	Frame Alignment Signal. (Used in the alignment of digital transmission frames.)
FAX	Facsimile.
FCC	Federal Communications Commission. (US authority, appointed by the President to regulate all interstate and international telecommunications.)
FCS	Frame Check Sequence. (Field added to a transmitted frame to check for errors.)
FDDI	Fibre Distributed Digital Interface. (Standard for optical fibre transmission.)
FDL	Facilities Data Link. (See EOC.)
FDM	Frequency Division Multiplexing. (Signal multiplexing technique.)
FDMA	Frequency Division Multiple Access. (Multiple access technique based on FDM.)
FDX	Full Duplex. (Transmission system in which the two stations connected by a link can transmit and receive simultaneously.)

FEC	Feedforward Error Correction. (Also called Forward Error Correction. Technique for correcting errors due to transmission.)
FEXT	Far End Crosstalk.
FFSK	Fast Frequency Shift Keying.
FHSS	Frequency Hopping Spread Spectrum.
FIFO	First In First Out. (Technique for buffering data.)
FM	Frequency Modulation. (Analogue signal modualtion technique.)
FMFB	Frequency Modulation Feedback.
FPODA	Fixed Priority Oriented Demand Assignment. (Medium multiple access method.)
FPS	Fast Packet Switch. (Standard for transmission based on frame relay or cell relay.
FSK	Frequency Shift Keying. (Digital modulation technique.)
FTAM	File Transfer and Access Method. (Or File Transfer Access and Management. International standard.)
FTP	File Transfer Protocol. (Used within TCP/IP.)
GDN	Government Data Network. (UK private data network for use by government departments.)
GDP	Gross Domestic Product. (Measure of output from a country.)
GEN	Global European Network. (Joint venture between European PTOs to provide high speed leased line and switched services. Likely to be replaced by METRAN in mid 1990s.)
GMSK	Gaussian Minimum Shift Keying. (Modulation technique, as used in GSM.)
GNMP	Government Network Management Profile. (Government procurement standard for network management, as part of GOSIP.)
GNP	Gross National Product.
GoS	Grade of Service. (Measure of service performance as perceived by the user.)

GOSIP	Government OSI Profile. (Government procurement standard.)
GVPN	Global Virtual Private Network.
HCI	Human Computer Interface.
HDB3	High Density Bipolar 3. (Line transmission encoding technique.)
HDLC	Higher level Data Link Control. (ITU-T bit oriented protocol for handling data.)
HDSL	High bit rate Digital Subscriber Line. (Bellcore technical advisory for the transmission of high bit rate data over twisted copper lines.)
HOMUX	Higher Order Multiplexer.
IA2	International Alphabet 2. (Code used in a teleprinter, also called the Murray code.)
IA5	International Alphabet 5. (International standard alphanumeric code, which has facility for national options. The US version is ASCII.)
IAB	Internet Activities Board.
IBC	Integrated Broadband Communications. (Part of the RACE programme.)
IBCN	Integrated Broadband Communications Network.
ICMP	Internet Control Message Protocol. (Protocol developed by DARPA as part of Internet for the host to communicate with gateways.)
IEC	International Electrotechnical Commission.
IEC	Interexchange Carrier. (US term for any telephone operator licensed to carry traffic between LATAs interstate or intrastate.)
IEEE	Institute of Electrical and Electronics Engineers. (USA professional organisation.)
IETF	Internet Engineering Task Force.
I-ETS	Interim European Telecommunicaitons Standard. (ETSI.)

IKBS	Intelligent Knowledge Based System.
IM	Intermodulation.
IN	Intelligent Network.
IPVC	International Private Virtual Circuit.
ISDN	Integrated Services Digital Network. (Technique for the simultaneous transmission of a range of services, such as voice, data and video, over telephone lines.)
ISLAN	Integrated Services Digital Network. (LAN which can carry an integrated service, such as voice, data and image.)
ISM	Industrial, Scientific and Medical. (Usually refers to ISM equipment or applications.)
ISO	International Standardisation Organization.
ISP	International Standardised Profile.
ITU	International Telecommunication Union.
ITU-T	International Telecommunication Union Telecommunication sector.
ITU-R	International Telecommunication Union Radiocommunication sector.
IVDS	Interactive Video and Data Services.
IVDT	Integrated Voice and Data Terminal. (Equipment with integrated computing and voice capabilities. In its simplest form it consists of a PC with telephone incorporated. Facilities such as storage and recall of telephone numbers is included.)
IXC	Interexchange Carrier. (USA long distance telecommunication carrier.)
IXI	International X.25 Infrastructure. (Pilot backbone pan-European network used by Europe's academic community.)
JEIDA	Japan Electronic Industry Development Association.
JISC	Japanese Industrial Standards Committee. (Standards making body which is funded by the Japanese government.)

JSA	Japanese Standards Association.
JTM	Job Transfer and Manipulation. (Communication protocols used to perform tasks in a network of interconnected open systems.)
LAN	Local Area Network. (A network shared by communicating devices, usually on a relatively small geographical area. Many techniques are used to allow each device to obtain use of the network.)
LAP	Link Access Protocol.
LAPB	Link Access Protocol Balanced. (X.25 protocol.)
LAPD	Link Access Protocol Digital. (ISDN standard.)
LAPM	Link Access Protocol for Modems. (ITU-T V.42 standard.)
LASER	Light Amplification by Stimulated Emission of Radiation. (Laser is also used to refer to a component.)
LATA	Local Access and Transport Area. (Area of responsibility of local carrier in USA. When telephone circuits have their start and finish points within a LATA they are the sole responsibility of the local telephone company concerned. When they cross a LATA's boundary, i.e. go inter-LATA, they are the responsibility of an interexchange carrier or IEC.)
LCN	Local Communications Network. (ITU-T)
LDM	Linear Delta Modulation. (Delta modulation technique in which a series of linear segments of constant slope provides the input time function.)
LEC	Local Exchange Carrier. (USA local telecommunication carrier.)
LLC	Logical Link Control. (IEEE 802. standard for LANs.)
LLP	Lightweight Presentation Protocol.
LPC	Linear Predictive Coding. (Encoding technique used in pulse code modulation.)
LRC	Longitudinal Redundancy Check. (Error checking procedure for transmitted data.)

LSB	Least Significant Bit. (Referring to bits in a data word.)
LTE	Line Terminating Equipment. (Also called Line Terminal Equipment. Equipment which terminates a transmission line.)
MA	Multiple Access.
MAC	Media Access Control. (IEEE standard 802. for access to LANs.)
MAN	Metropolitan Area Network.
MAP	Manufacturing Automation Protocol.
MAT	Metropolitan Area Trunk. (A cable system which is used to reduce crosstalk effects in regions where there is a large number of circuits between exchanges.)
MDNS	Managed Data Network Service. (Earlier proposal by CEPT which has now been discontinued.)
METRAN	Managed European Transmission Network. (CEPT initiative to provide a broadband backbone across Europe.)
MFJ	Modification of Final Judgement. (Delivered by Judge Harold Greene. 1982 act in the AT&T divestiture case.)
MHS	Message Handling System. (International standard.)
MITI	Ministry of International Trade and Industry. (Japanese.)
MLM	Multi-level Multi-Access protocol. (Multiple access technique.)
MMI	Man Machine Interface. (Another name for the human-computer interface or HCI.)
MoD	Ministry of Defence. (UK)
MODEM	Modulator/Demodulator. Device for enabling digital data to be send over analogue lines.
MoU	Memorandum of Understanding.
MPT	Ministry of Posts and Telecommunications. (Japan.)

MSAP	Mini Slotted Alternating Priorities. (Multiple access technique.)
MSK	Minimum Shift Keying. (A form of frequency shift keying, or FSK.)
MTN	Managed Transmission Network.
NAFTA	North American Free Trade Association.
NAK	Negative Acknowledgement. (In data transmission this is the message sent by the receiver to the sender to indicate that the previous message contained an error, and requesting a re-send.)
NANP	North American Numbering Plan. (Telephone numbering scheme administered by Bellcore.)
NBS	National Bureau of Standards. (USA.)
NCL	Network Control Layer.
NCP	Network Control Point.
NE	Network Element.
NET	Nome Europeenne de Telecommunication. (European Telecommunications Standard, which is mandatory.)
NEXT	Near End crosstalk. (The unwanted transfer of signal energy from one link to another, often closely located, at the end of the cable where the transmitter is located.)
N-ISDN	Narrowband Integrated Services Digital Network.
NIST	National Institute for Standards and Technology. (USA.)
NLC	Network Level Control. (Or Network Layer Control.)
NMC	Network Management Centre.
NMI	Network Management Interface. (Term used within OSI to indicate the interface between the network management system and the network it manages.)
NNI	Network Node Interface. (Usually the internal interfaces within a network. See UNI.)

NRZ	Non Return to Zero. (A binary encoding technique for transmission of data.)
NRZI	Non Return to Zero Inverted. (A binary encoding techniqe for transmission of data.)
NSAP	Network Service Access Point. (Prime address point used within OSI.)
NT	Network Termination. (Termination designed within ISDN e.g. NT1 and NT2.)
NTE	Network Terminating Equipment. (Usually refers to the customer termination for an ISDN line.)
NTIA	National Telecommunications Industry Administration. (USA)
NTN	Network Terminal Number. (Part of an international telephone number.)
OA&M	Operations, Administration and Maintenance. (Also written as OAM.)
OAM&P	Operations, Administration, Maintenance & Provisioning.
OCDMA	Orthogonal Code Division Multiple Access.
ODA	Open Document Architecture. (Or Office Document Architecture. Standard for transmission of content and layout of a document and of multimedia documents.)
ODIF	Office Document Interchange Format. (Format used to communicate documents within an open system.)
ODP	Open Distributed Processing.
OECD	Organisation for Economic Co-operation and Development.
OEM	Original Equipment Manufacturer. Supplier who makes equipment for sale by a third party. The equipment is usually disguised by the third party with his own labels.)
OFTEL	Office of Telecommunications. (UK regulatory body.)

OIW	OSI Implementors' Workshop. (USA based.)
OKQPSK	Offset Keyed Quaternary Phase Shift Keying.
O&M	Operations & Maintenance.
ONA	Open Network Architecture.
OND	Open Network Doctrine. (Japanese plan for equal access.)
ONP	Open Network Provision.
OOK	On-Off Keying. (Digital modulation technique. Also known as ASK or Amplitude Shift Keying.)
OOP	Object Oriented Programming.
OS	Operating System. (Or Operations System. ITU-T.)
OSF	Open Software Foundation
OSF	Operations System Function. (ITU-T.)
OSI	Open Systems Interconnection. (Refers to the seven layer reference model.)
OSIE	OSI Environment.
OSITOP	Open Systems Interconnection Technical and Office Protocol.
PAD	Packet Assembler/Disassembler. (Protocol converter used to provide access into the packet switched network.)
PAM	Pulse Amplitude Modulation. (An analogue modulation technique.)
PC	Private Circuit.
PCM	Pulse Code Modulation. (Transmission technique for digital signals.)
PDH	Plesiochronous Digital Hierarchy. (Plesiochronous transmission standard.)
PDM	Pulse Duration Modulation. (Signal modulation technique, also known as Pulse Width Modulation or PWM.)
PDU	Protocol Data Unit. (Data and control information passed between layers in the OSI Seven Layer model.)
PFM	Pulse Frequency Modulation. (An analogue modulation technique.)

PLP	Packet Level Protocol.
PM	Phase Modulation. (Analogue signal modulation technique.)
PODA	Priority-Oriented Demand Assignment protocol. (Multiple access technique. See also FPODA and CPODA.)
POSI	Promoting conference for OSI. (Japan and Far East users' group active in functional standards and interconnection testing.)
POTS	Plain Old Telephone Service. (A term loosely applied to an ordinary voice telephone service.)
PPM	Pulse Phase Modulation. (An analogue modulation technique. Sometimes called Pulse Position Modulation.)
PRA	Primary Rate Access. (ISDN, 30B+D or 23B+D code.)
PRBS	Pseudo Random Binary Sequence. (Signal used for telecommunication system testing.)
PRF	Pulse Repetition Frequency. (Of a pulse train.)
PRK	Phase Reversal Keying. (A modification to the PSK modulation technique.)
PRMA	Packet Reservation Multiple Access.
PSDN	Packet Switched Data Network. (Or Public Switched Data Network. X.25 network, which may be private or public.)
PSPDN	Packet Switched Public Data Network.
PSE	Packet Switching Exchange.
PSK	Phase Shift Keying. (Analogue phase modulation technique.)
PSN	Packet Switched Network.
PSN	Public Switched Network.
PSS	Packet Switched Service. (Data service offered by BT.)
PSTN	Public Switched Telephone Network. (Term used to describe the public dial up voice telephone network, operated by a PTT.)
PTN	Public Telecommunications Network.

PTO	Public Telecommunication Operator. (A licensed telecommunication operator. Usually used to refer to a PTT.)
PTT	Postal, Telegraph and Telephone. (Usually refers to the telephone authority within a country, often a publicly owned body. The term is also loosely used to describe any large telecommunications carrier.)
PUC	Public Utility Commission. (In USA.)
PVC	Permanent Virtual Circuit. (Method for establishing a virtual circuit link between two nominated points. See also SVC)
PWM	Pulse Width Modulation. (Analogue modulation technique in which the width of pulses is varied. Also called Pulse Duration Modulation, PDM, or Pulse Length Modulation, PLM.)

QAM	Quadrature Amplitude Modulation. (A modulation technique which varies the amplitude of the signal. Used in dial up modems. Also known as Quadrature Sideband Amplitude Modulation or QSAM.)
QD	Quantising Distortion.
QoS	Quality of Service. (Measure of service performance as perceived by the user.)
QPRS	Quadrature Partial Response System. (Signal modulation technique.)
QPSK	Quadrature Phase Shift Keying. (Signal modulation technique.)

RBER	Residual Bit Error Ratio. (Measure of transmission quality. ITU-T Rec. 594-1.)
RBOC	Regional Bell Operating Company. (US local carriers formed after the divestiture of AT&T.)
RFI	Radio Frequency Interference.
RFNM	Ready For Next Message.

RTS   Request To Send. (Handshaking routine used in anlogue transmission, such as by modems.)

RZ   Return to Zero. (A digital transmission system in which the binary pulse always returns to zero after each bit.)

SAP   Service Access Point. (Port between layers in the OSI seven layer model, one for each of the layers, e.g. LSAP, NSAP, etc.)

SAPI   Service Access Point Identifier. (Used within ISDN Layer 2 frame.)

SDH   Synchronous Digital Hierarchy.

SDU   Service Data Unit. (Data passed between layers in the OSI Seven Layer model.)

SED   Single Error Detecting Code. (Transmission code used for detecting errors by use of single parity checks.)

SELV   Safety Extra Low Voltage circuit. (A circuit which is protected from hazardous voltages.)

SIO   Scientific and Industrial Organisation.

SITA   Societe Internationale de Telecommunications Aeronautiques. (Refers to the organisation and its telecommunication network which is used by many of the world's airlines and their agents, mainly for flight bookings.)

SMDS   Switched Multimegabit Data Service. (High speed packet based standard proposed by Bellcore.)

SMF   System Management Function.

SNAP   Subnetwork Access Protocol. (IEEE protocol which allows non-OSI protocols to be carried within OSI protocols.)

SNI   Subscriber Network Interface.

SNMP   Simple Network Management Protocol. (Network management system within TCP/IP.)

SNR   Signal to Noise Ratio.

SOHO   Small Office Home Office (market).

SONET	Synchronous Optical Network. (Synchronous optical transmission system developed in North America, and which has been developed by ITU-T into SDH.)
SP	Service Provider.
SPADE	Single channel per carrier Pulse code modulation multiple Access Demand assignment Equipment.
SQNR	Signal Quantisation Noise Ratio.
SSB	Single Sideband.
SSBSC	Single Sideband Suppressed Carrier modulation. (A method for amplitude modulation of a signal.)
SSMA	Spread Spectrum Multiple Access.
STDM	Synchronous Time Division Multiplexing.
STDMX	Statistical Time Division Multiplexing.
STE	Signalling Terminal Equipment.
STM	Synchronous Transport Module. (Basic carrier module used within SDH, e.g. STM-1, STM-4 and STM-16.)
STMR	Sidetone Masking Rating. (Measure of talker effects of sidetone.)
STP	Shielded Twisted Pair. (Cable.)
STP	Signal Transfer Point.
STS	Synchronous Transport Signal. (SONET standard for electrical signals, e.g. STS-1 at 51.84Mbit/s.)
STX	Start of Text. (Control character used to indicate the start of data transmission. It is completed by a End of Text character, or ETX.)
SVC	Switched Virtual Circuit. (Method for establishing any to any virtual circuit link. See also PVC.)
TA	Terminal Adaptor. (Used within ISDN to convert between non-ISDN and ISDN references.)
TA	Telecommunication Authority.
TA 84	Telecommunications Act of 1984. (UK.)
TACS	Total Access Communication Systems. (Adaption of AMPS by the UK to suit European frequency allocations.)

TAPI	Telephony Applications Programming Interface. (A standard for linking telephones to PCs.)
TASI	Time Assignment Speech Interpolation. (Method used in analogue speech transmission where the channel is activated only when speech is present. This allows several users to share a common channel.)
TC	Transport Class. (e.g. TC 0, TC 4, etc.)
TCM	Time Compression Multiplexing. (Technique which separates the two directions of transmission in time.)
TCM/DPSK	Trellis Coded Modulation/Differential Phase Shift Keying.
TCP/IP	Transmission Control Protocol/ Internet Protocol. (Widely used transmission protocol, originating from the US ARPA defence project.)
TDD	Time Division Duplex transmission.
TDD/FDMA	Time Division Duplex/Frequency Division Multiple Access. (Means of multiplexing several two way calls using many frequencies, with a single two way call per frequency.)
TDD/TDMA	Time Division Duplex/Time Division Multiple Access. (Means of multiplexing two way calls using a single frequency for each call and multiple time slots.)
TDM	Time Division Multiplexing. (Technique for combining, by interleaving, several channels of data onto a common channel. The equipment which does this is called a Time Division Multiplexer.)
TDMA	Time Division Multiple Access. (A multiplexing technique where users gain access to a common channel on a time allocation basis. Commonly used in satellite systems, where several earth stations have total use of the transponder's power and bandwidth for a short period, and transmit in bursts of data.)
TE	Terminal Equipment.

TEI	Terminal Endpoint Identifier. (Used within ISDN Layer 2 frame.)
TEMA	Telecommunication Equipment Manufacturers Association. (UK.)
TIA	Telecommunication Industry Association. (US based. Formed from merger of telecommunication sector of the EIA and the USTSA.)
TMA	Telecommunication Managers' Association. (UK)
TMN	Telecommunications Management Network.
TR	Technical Report. (ISO technical document; not a standard.)
TTE	Telecommunication Terminal Equipment.
TTY	Teletypewriter. (Usually refers to the transmission from a teletypewriter, which is asynchronous ASCII coded.)
TUA	Telecommunications Users' Association. (UK.)
TWX	Teletypewriter exchange service. (Used in Canada.)
UAP	User Application Process.
UART	Universal Asynchronous Receiver/Transmitter. (The device, usually an integrated circuit, for transmission of asynchronous data. See also USRT and USART.)
UDF	Unshielded twisted pair Development Forum. (Association of suppliers promoting transmission over UTP.)
UI	User Interface.
UL	Underwriters Laboratories. (Independent USA organisation involved in standards and certification.)
UNI	User Network Interface. (Also called User Node Interface. External interface of a network.)
UPS	Uninterrupted Power Supply. (Used where loss of power, even for a short time, cannot be tolerated.)
USART	Universal Synchronous/Asynchronous Receiver/Transmitter. (A device, usually an integrated circuit, used in data communication devices, for

	conversion of data from parallel to serial form for transmission.)
USB	Upper Sideband.
USRT	Universal Synchronous Receiver/Transmitter. (A device, usually an integrated circuit, which converts data for transmission over a synchronous channel.)
UTP	Unshielded Twisted Pair. (Cable.)

VADS	Value Added Data Service.
VAN	Value Added Network
VANS	Value Added Network Services.
VAS	Value Added Service. (See also VANS.)
VASP	Value Added Service Provider.
VBR	Variable Bit Rate.
VPN	Virtual Private Network. (Part of a network operated by a public telephone operator, which is used as a private network.)
VRC	Vertical Redundancy Check. (Parity method used on transmitted data for error checking.)
VSB	Vestigial Sideband modulation. (A method for amplitude modulation of a signal.)

WACK	Wait Acknowledgement. (Control signal returned by receiver to indicate to the sender that it is temporarily unable to accept any more data.)
WAN	Wide Area Network.

Index